D1525894

FIRE POWER IN THE LAB

Automation in the Fight Against Infectious Diseases and Bioterrorism

Scott P. Layne,
Tony J. Beugelsdijk,
and
C. Kumar N. Patel,
Editors

JOSEPH HENRY PRESS
Washington, D.C.

ST. PHILIP'S COLLEGE LIBRARY

R
858
.A2
F55
2001
Joseph Henry Press • 2101 Constitution Avenue, N.W. • Washington, D.C. 20418

The Joseph Henry Press, an imprint of the National Academy Press, was created with the goal of making books on science, technology, and health more widely available to professionals and the public. Joseph Henry was one of the founders of the National Academy of Sciences and a leader of early American science.

This volume is based on a colloquium "Automation in Threat Reduction and Infectious Disease Research: Needs and New Directions," held in Washington, D.C., on April 29-30, 1999. Financial support for this project was provided by the Centers for Disease Control and Prevention, the U.S. Department of Energy, the U.S. Department of Health and Human Services Office of Emergency Preparedness, and the Los Alamos National Laboratory. Additional sponsorship was provided by the Association for Laboratory Automation, the Institute of Medicine, the National Academy of Engineering, and the University of California, Los Angeles.

The contents of this volume are based on presentations and discussions that took place during the colloquium. Authors had the opportunity to update references up to the date of publication. Any opinions, findings, conclusions, or recommendations expressed in this volume are those of the editors or authors and do not necessarily reflect the views of the National Academies or the organizations that provided support for the project.

Library of Congress Cataloging-in-Publication Data

Firepower in the lab : automation in the fight against infectious diseases and bioterrorism / Scott P. Layne, Tony J. Beugelsdijk, and C. Kumar N. Patel, editors.
p. ; cm.
Includes bibliographical references and index.
ISBN 0-309-06849-5 (alk. paper)
1. Medical laboratories—Automation—Congresses. 2. Medicine—Research—United States. 3. Laboratories—Technological innovations—United States. 4. Bioterrorism. I. Layne, Scott P. II. Beugelsdijk, Tony J., 1949- III. Patel, C. Kumar N.
[DNLM: 1. Automation—Congresses. 2. Communicable Disease Control—Congresses. 3. Containment of Biohazards—Congresses. 4. Food Contamination—prevention & control—Congresses. 5. Laboratory Infection—prevention & control—Congresses. QY 23 F523 2001]
R858.A2 F55 2001
610'.285—dc21 2001024533

Copyright 2001 by the National Academy of Sciences. All rights reserved.

Printed in the United States of America.

CONTENTS

PREFACE

In April 1999 a group of 200 experts in the fields of infectious diseases, food safety, bioterrorism mitigation, and molecular medicine gathered for a colloquium at the National Academy of Sciences in Washington, D.C. What brought this diverse group of individuals together was the understanding that each of their various disciplines must contend with certain "grand" problems that require enormous quantities of laboratory-based data to make progress. The various scientific challenges discussed during the meeting included (1) fighting deadly infectious diseases such as influenza A epidemics, multidrug-resistant tuberculosis, and the human immunodeficiency virus; (2) ensuring safe food by reducing risks, detecting pathogens and toxins, and investigating infectious disease outbreaks; (3) mitigating bioterrorism and biowarfare by preventing attacks, characterizing agents, and minimizing aftermaths; and (4) facilitating work on human genetics and molecular medicine, especially in predicting cancers, diagnosing diseases, and tailoring medications.

Meeting participants identified the various needs in their disciplines as well as the accomplishments that would be possible if such needs were met with high-throughput laboratory and informatics resources. Participants emphasized problems involving large populations of people and/or significant numbers of genetic variations that, because of their sheer size, justify focused efforts to create and analyze enormous quantities of laboratory-based data. Within several years the associated databases are envisioned to grow to $\sim 10^{15}$ bits (petabits); thus, the limiting factor is that humans, unaided, are not capable of producing such vast inventories of data.

Specialists surveyed the available building blocks for establishing petabit-generating user facilities—including robotics, laboratory automation, lab-on-a-chip, informatics, Internet, and process control innovations. They noted that, for the first time, all of the necessary pieces are available for building flexible and programmable laboratory firepower (a capability referred to as "mass customized testing via the Internet"). Although petabit-generating labs are available to medical and biological researchers in industry, they are not available to researchers in academic institutions. Such allocation of resources is currently shifting large-scale undertakings into the private sector and away from the public domain.

Policymakers noted that few, if any, government grants are available for intermediate-cost research efforts (e.g., user facilities), although grants are available for many small-cost (e.g., individual-investigator) and certain large-cost projects (e.g., human genome). Herein lies the most significant obstacle to obtaining adequate funding. User facilities that are dedicated to tackling grand problems will likely cost $20 million to start up over a 2-year period and $5 million to supply and operate annually. Yet no single government agency offers grants for such intermediate-cost research efforts, and frequently it is impossible to coordinate major funding across two or more agencies. Major philanthropic foundations may be a more realistic source of funds.

Meeting participants also discussed what approaches might be effective in surmounting such funding obstacles. One practical idea that gained wide support at the colloquium was that if researchers start with an especially compelling problem—such as a widespread and potentially catastrophic infectious disease—government agencies ultimately might be persuaded to offer much needed medium-level grants.

The chapters in this volume come from invited talks at the colloquium. They are arranged according to the four scientific challenges—infectious diseases, food safety, bioterrorism, and genomics—discussed during the meeting, with more detailed "technology" chapters interspersed between "scientific" chapters. The second chapter, "Tackling Grand Challenges with Powerful Technologies" offers an overview and perspective on the key topics in this volume. The last chapter, "Summary and Next Steps," offers unofficial, yet important, conclusions and recommendations that emerged from the meeting.

We wish to thank the colloquium's sponsors for institutional and financial support: Association for Laboratory Automation; Centers for Disease Control and Prevention; U.S. Department of Energy; U.S. Department of Health and Human Services Office of Emergency Preparedness; Institute of Medicine; Los Alamos National Laboratory; National Academy of Engineering; and the University of California, Los Angeles. We also thank the members of the organizing committee for their help in building the program: Donald S. Burke, Irvin S. Y. Chen, Raymond E. Dessy, Robin A. Felder, Maurice Hilleman, James M. Hughes, Gary W. Kramer, David J. Lipman, Michael Osterholm, William E. Paul, Robert T. Schooley, and Alejandro C. Zaffaroni. We especially thank Jonathan Davis and Andrew Pope at the Institute of Medicine for their interests and extra efforts. Finally, we acknowledge and thank Kathi E. Hanna for her expert editorial assistance in preparing this volume.

SCOTT P. LAYNE,
TONY J. BEUGELSDIJK, and
C. KUMAR N. PATEL, *Editors*

FIRE**POWER** IN THE **LAB**

1

Where the Bucks Stop: A Case for Intermediate-Scale Grants

C. Kumar N. Patel

Both the pace and the scale of research are increasing in all sciences, especially the biological and medical disciplines. There are two forces at work. First, humans around the world face "grand challenges" from infectious diseases that involve large populations of people and/or significant numbers of genetic variations. Because of their sheer size and risk, such problems justify focused efforts to create and analyze enormous quantities of laboratory-based data. Second, the very nature of many of these diseases requires capabilities for rapid response from the research and clinical communities. The long-term societal costs for not being able to respond to these challenges are likely to be high. As a response to the perceived needs, relatively small efforts (composed of a few technicians working at the bench) are giving way to much larger automated laboratory efforts that analyze and generate quantities of information often amounting to petabytes (10^{15}). Leading universities around the country, having seen what lies ahead, are struggling to respond by initiating high-throughput research and development programs for competing in the "post-genomics" era. The start-up costs for such programs can easily exceed $25 million over two to three years. The problem is that few, if any, government grants are available for intermediate-cost (e.g., user-centric as opposed to facilities-centric) research efforts. On the other hand, we have done well in the areas of small "single private-investigator" grants (PI-centric) and very large grants, which are directed toward construction, and in the operation of "one-of-a-kind" national efforts (facility-centric national efforts as characterized by synchrotron and neutron facilities, which are key for many biological studies).

Small-cost grants are plentiful and typified by RO1 awards from the National Institutes of Health (NIH). Generally speaking, such PI-centric awards provide less than $2 million dollars, span two to four years, and support one to three investigators. Large grants are not plentiful, but nonetheless allocate substantial amounts of national research expenditures and provide facilities support to a large number of PIs on a visiting basis. In contrast, intermediate-cost grants are relatively few and far between, sometimes approved in exceptional circumstances but never really sustained as a mechanism that meets a national need. Such user-centric awards would provide $25 million dollars, span four to five years, and support 10 to 15 investigators. The key difference between the traditionally large facilities-centric programs and the proposed user-centric programs is that the latter can be cloned and adapted to similar types of problems at many locations throughout the country. In this model the first-generation facility serves as a prototype or shakedown, with second-generation facilities costing far less. A cutting-edge program at one institution can thereby seed innovative programs at others.

Large-cost grants are represented by line-item expenditures in congressional budgets and are targeted to specific government agencies. Such funding arises from efforts drawn out over extended periods of time and involving scientific consensus, congressional testimony, and commissioned studies. Such facility-centric awards often provide over $50 million, have buildup times of three to five years before becoming operational, support teams of many simultaneous users, and are national in scope. Because of its size and operational complexity, the one-of-a-kind facility cannot be physically moved or easily cloned. Various Department of Energy (DOE) -based synchrotron and neutron sources have such features; so does the NIH-based National Human Genome Research Institute.

It is clear that many of the problems discussed during the colloquium "Automation in Threat Reduction and Infectious Disease Research: Needs and New Directions" fall into the category of intermediate-cost grants. It also is clear that building new automated high-throughput laboratories is scientifically and technically feasible. Such intermediate-scale resources would take advantage of existing know-how, would be operational within one to two years, and would enable new solutions to grand challenges. There is a clear and present national need to address the policy vacuum for providing such intermediate-scale grants. That such a hole exists in the national granting process is clearly evidenced from data gathered from NIH, DOE, and the Department of Defense (DOD).

I believe the topics covered in this volume will give elected representatives and high-level federal officials sufficient ammunition to move forward. Most of the infectious disease, food safety, and bioterrorism-

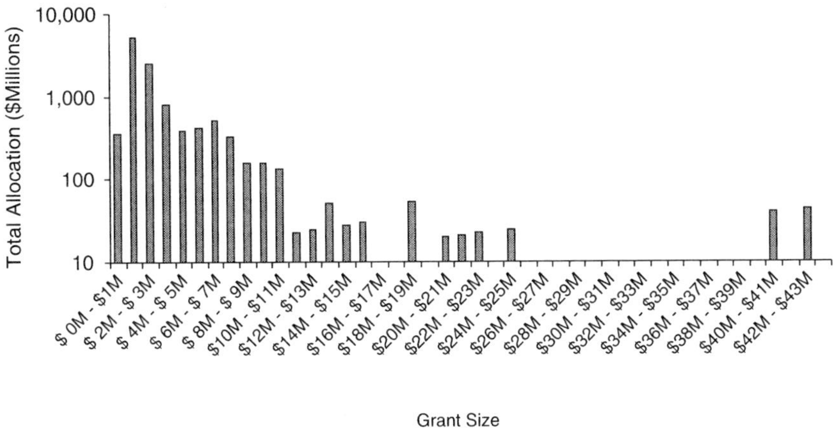

FIGURE 1.1 Distribution of NIH funds by grant size (average duration four years).

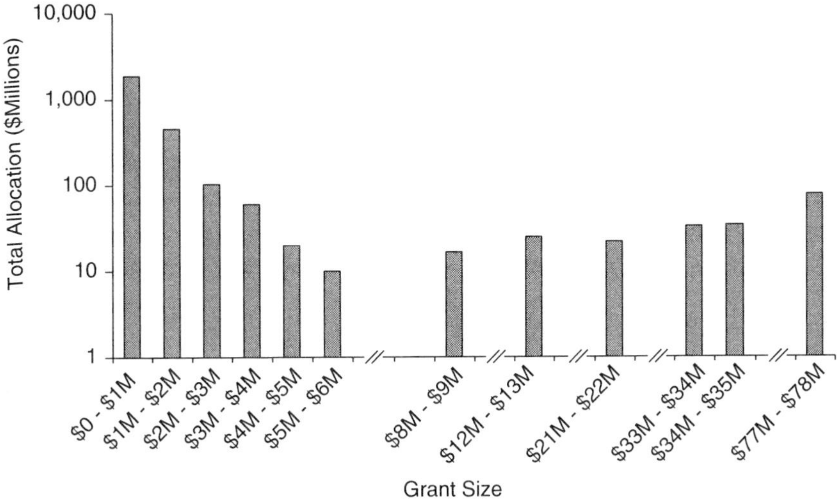

FIGURE 1.2 Distribution of National Science Foundation (NSF) funds by grant size (average duration four years).

ST. PHILIP'S COLLEGE LIBRARY

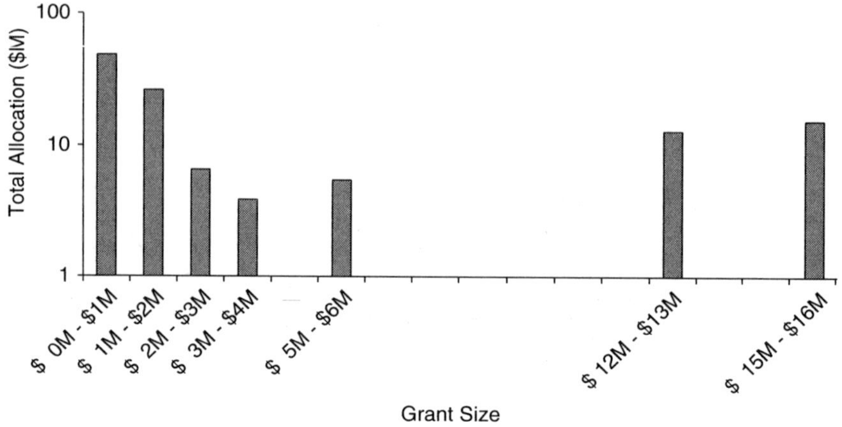

FIGURE 1.3 Distribution of DOE funds by grant size (average duration three years).

related applications discussed are for the public good and therefore not in direct competition with the private sector. With the current wave of economic prosperity and sharp reductions in the federal deficit, the time is doubly right to facilitate intermediate-cost grants.

I would choose compelling new directions that serve as test cases and within the next one to two years would allocate up to 10 intermediate-cost grants at $25 million each. Appropriate national committees such as those supported by the National Research Council should evaluate the outcome and effectiveness of such grants, since this funding mode is a departure from the business-as-usual mode. If a thorough review confirms that such grants have catalyzed tangible benefits, a strong case can be made for continuation of intermediate-cost grants. The $125 million to $250 million expenditure would then be a wise investment and a sound experiment in scientific policy.

2

Tackling Grand Challenges with Powerful Technologies

Scott P. Layne, Tony J. Beugelsdijk, and C. Kumar N. Patel

Headline news stories about illnesses and deaths in Hong Kong caused by avian influenza viruses, multidrug resistant *Mycobacterium tuberculosis,* and foodborne Salmonella, plus new threats from bioterrorists, have become commonplace in recent years. Such headlines are reminders that we are not immune to the dangers of infectious diseases just because we have been immunized, consume regulated food and water, and are protected by modern systems of public health and national security. Through concerted efforts we have eradicated or virtually eliminated certain infectious diseases that once devastated the world's population (e.g., smallpox and polio), but our past achievements in these areas—plus the widespread availability of lifesaving antibiotics—have led to complacency. In fact, we are engaged in ongoing contests with no clear end in sight against infectious diseases, foodborne pathogens, and international terrorism. These battles have many common elements that require a coordinated armamentarium from scientific research, modern technology, and public health preparedness.

In comparison to other weapons of mass destruction (i.e., chemical and nuclear), biological weapons are relatively simple and inexpensive to develop. Today, many experts believe it is no longer a question of if but rather when bioweapons will be unleashed by rogue states or terrorist groups. Small attacks may strike as few as 10 to 100 people yet may lead to lasting changes in the fabric of civilized societies. Thus, in order to deal with the looming threat of bioterrorist attacks, public health, law enforcement, and military-based laboratories are now preparing to identify pathogens that are seldom seen in patients and would likely go unrecognized if the diagnosis depended on initial clinical signs and symptoms.

Within the next few years, scientists will sequence the entire human genome for a dozen or so people. Such information will move the practice of medicine in the direction of prevention and treatment based on molecular makeups. Initially, discovering how health (epidemiologic outcomes) is dependent on single mutations (genotypic features) and single proteins (phenotypic features) will be relatively straightforward. For example, it may involve understanding how single nucleotide polymorphisms (SNPs) are related to developing cancers, diagnosing diseases, and tailoring medications. With time, however, more intricate relationships likely will be discovered with population-based studies, entailing the generation, storage, and analysis of enormous quantities of epidemiologic, genotypic, and phenotypic data.

On the "digital" side, researchers have growing access to computers that perform 10^{12} floating point operations per second (teraflops), Internet backbones that transmit 10^9 bits of information per second (gigahertz), and databases that contain 10^{15} bits of information (petabits). On the "physical" side, researchers must still depend on graduate students, postdoctoral fellows, and technicians to perform many repetitious tests. In yesterday's model of research and discovery, such manual approaches were sufficient. In today's model, however, such "digital" versus "physical" mismatches are constraining the exploration of enormous experimental and informatics relationships (or phase spaces) in medicine and biology.

Thus, more than ever, new tools and technologies from engineering, computer science, mathematics, chemistry, and physics hold great promise for tackling "grand" problems in medicine and biology. The challenge is to identify important scientific needs and then focus on the current technologies and approaches that are available for speeding up the pace of laboratory-based research. Given such perspectives, the colloquium considered specific needs in medicine and biology, assessed current research practices and their limitations, and then considered strategic ways for developing new high-throughput laboratories and informatic resources. This kind of multidisciplinary approach attracted broad-based participation from academic, governmental, and industrial sectors.

GRAND CHALLENGE PROBLEMS FOR SCIENCE

The triple threats of communicable infectious diseases, pathogenic foodborne infections, and bioterrorism/biowarfare pose enormous challenges to established research and public health infrastructures. Such challenges share a common set of themes: (1) they affect large numbers of people; (2) they require large quantities of data to achieve solutions;

(3) they require access to appropriate and often large numbers of samples for testing and analysis; (4) they demand reproducible laboratory procedures; and (5) they require sufficient quantities of supporting reagents. In today's scientific research and public health laboratories, most activities are carried out by technicians using various labor-saving devices that are "semiautomated." Although such approaches have led to insights and breakthroughs, they also demand the completion of highly repetitive and tedious tasks by armies of laboratory technicians.

The model of semiautomated/semimanual laboratories has three significant shortcomings. First, because such facilities require significant human involvement, they scale linearly in cost with the overall size of the problem. Second, because such facilities demand humans to perform highly repetitive and tedious tasks, they are prone to random errors. Third, because such facilities inevitably contend with rate-limiting tasks, they can be overwhelmed by large surges in demand. Such acute surges in demand may occur, for example, during a rapidly growing epidemic or bioterrorist attacks, where swift diagnosis and follow-up are necessary for an effective response.

Fortunately, a critical number of scientific disciplines and powerful technologies can be combined to level the playing field against the triple threats outlined above. Molecular biologists and biochemists have developed a variety of laboratory-based assays that are powerful and readily adaptable to large-scale efforts. Engineers have developed innovative robotics and automation technologies that are capable of skyrocketing the number and variety of laboratory experiments. Computer scientists have developed programming languages and database management systems that have provided the basic building blocks for improved environments for scientific collaboration. And physicists—driven by the need to share large amounts of data generated in high-energy particle physics—have catalyzed the development of the Internet, which is literally transforming the ways in which scientific and technical collaborations take place.

Colloquium participants recognized that integrating such disciplines and technologies holds great promise for accelerating infectious disease research—including basic science, clinical trials, and public health and epidemiologic investigations worldwide. In the coming years, naturally occurring and maliciously initiated outbreaks could be met with fast and accurate automated systems that facilitate sample collection, sample testing, data storage, and informatics analysis. Such automated systems could be flexible, modular, and remotely accessible via the Internet, thereby enabling a convenient means of mass customized testing.

FIGHTING ESTABLISHED AND EMERGING INFECTIOUS DISEASES

In the twenty-first century, infectious diseases will pose major challenges from a variety of sources. For example, new and lethal strains of influenza A H5N1 have surfaced in Hong Kong, which reinforces the need for an expanded program of global surveillance by the World Health Organization (WHO). New epidemics of multidrug-resistant *Mycobacterium tuberculosis* (MDR-TB) are emerging in countries belonging to the former Soviet Union, and the ability to administer the right combination of directly observed therapies (DOT) may depend on large-scale drug sensitivity testing. Worldwide, human immunodeficiency virus (HIV) infections and acquired immune deficiency syndrome (AIDS) cases are increasing exponentially, and discovering new drug therapies and vaccines may ultimately depend on brute force research and development efforts. As further described below, colloquium participants identified critical needs and limiting factors for each of these infectious diseases. They also considered ways to build high-throughput laboratory and informatics resources for accelerating basic science, clinical trials, and public health/epidemiologic investigations throughout the world.

Influenza

Influenza pandemics have swept the world for centuries, three times in the twentieth century alone (1918, 1957, and 1968). The 1918 influenza A H1N1 pandemic is the biggest infectious disease catastrophe on record, topping even the medieval Black Death in the number of deaths. Within months following the initial outbreak, an estimated 500 million people were infected and 40 million were dead from a worldwide population of 2 billion people.

In the United States the Centers for Disease Control and Prevention (CDC) has estimated that the next influenza pandemic may cause 89,000 to 207,000 deaths, 314,000 to 734,000 hospitalizations, 18 million to 42 million outpatient visits, and 20 million to 47 million additional illnesses. The overall economic impact may range from $71 billion to $167 billion dollars, excluding disruptions to commerce and society. Yet an often overlooked statistic is that the cumulative impact from numerous smaller epidemics is even greater than the toll from the recent pandemics combined. In "milder" epidemics, 10 percent of the world's population may be afflicted, which results in an estimated 500,000 annual deaths directly from influenza and its related complications. In "harsher" epidemics these grim worldwide statistics may readily double to 20 percent infected and 1 million annual influenza-related deaths.

At present, it is unclear whether the extreme virulence of the 1918 strain is a one-time event, perhaps due to some combination of unlikely factors, or whether similar catastrophes are forthcoming. Whatever the mix of factors, it is becoming increasingly clear that deciphering the 1918 strain will demand far more organized information on influenza than currently exists. New data are needed that correlate epidemiologic, phenotypic, and genotypic features of influenza viruses from around the world. High-throughput automated laboratories and informatics resources could transform the approach to deciphering the 1918 strain and others, like the avian influenza virus that struck residents in Hong Kong (see below).

Tuberculosis

One-third of the world's population has been exposed to *Mycobacterium tuberculosis*, resulting in latent infections that may reactivate at any point in a person's lifetime. As a consequence, there are now 16 million people with active cases of tuberculosis and 2 million deaths per year globally. Tuberculosis can be effectively treated with two to four first-line drugs (e.g., isoniazid, rifampin, ethambutol, and pyrazinamide) over several months but serious problems may arise when strains develop resistance to one or more of these agents.

Recent worldwide surveys indicate that a variety of MDR-TB strains are dispersed throughout the world and are apparently increasing in countries with strained public health infrastructures (Pablos-Mendez et al., 1998). In the United States, MDR-TB strains are especially prevalent among HIV-positive people in New York City, and infections with such highly resistant strains carry high mortality rates (even in otherwise healthy persons). In the former Soviet Union the emerging MDR-TB epidemic may involve as many as 11,400 cases, with the majority occurring in prisoners. In the coming years a public health disaster may arise from newly released prisoners transferring numerous drug-resistant strains to the general population.

In the United States, treating one MDR-TB patient with DOT may cost as much as $250,000 annually. For the world's public health system, which expends an average of $5 per person each year, such costs are staggering. Preventing the emergence and spread of MDR-TB is a global priority. Cost-effective means are needed to test for MDR-TB strains in large numbers of afflicted and/or exposed people throughout the world, and high-throughput automated laboratories could transform the very approach to diagnosing resistant strains. Such testing resources could rapidly identify optimal therapies for individuals and, because many second-line drugs are in scare supply, enable pharmaceutical companies to plan for future worldwide demands.

HIV/AIDS

In less than three decades HIV has grown from an unknown pathogen to a pandemic disease, chronically infecting 50 million people globally and killing 3 million persons each year. If such trends continue, AIDS will likely become the number one infectious disease killer in the world. In response to this catastrophe, pharmaceutical companies are manufacturing new antiviral drugs (e.g., reverse transcriptase and protease inhibitors) that extend lives, and investigators are examining promising leads for newer drugs and much-needed vaccines. Despite such progress, strains that are resistant to one or more antiviral drugs are becoming increasingly common, and reports of multidrug-resistant strains of HIV (MDR-HIV) being transmitted from person to person are now appearing. Given the cost of combination antiviral therapies and mounting problems with resistance, the only effective solution for the world will be the delivery of affordable vaccines that prevent new infections.

A simple estimate leads to the realization that there could be an enormous number of unique HIV strains (i.e., millions to billions) infecting people throughout the world. At present there is only sketchy information on these genetic mutations and their immunological and physiological correlates, making it difficult to relate the significance of such variations (if any) to therapies and vaccine development. High-throughput automated laboratories would enable investigators to conceptualize and attack the challenges pertaining to HIV from entirely new directions. For many such problems with enormous phase spaces, high-throughput laboratories that serve as investigational resources are perhaps the only feasible means to move ahead.

HIGH-THROUGHPUT AUTOMATED INFLUENZA LABORATORY

Fifty years ago the WHO established a global influenza surveillance network. Today, the network has grown to include 110 national collaborating laboratories in 80 countries worldwide. In the United States the domestic network includes over 70 collaborating laboratories, with many located in state health departments. Collaborating laboratories focus on collecting influenza samples from stricken people and verifying that samples contain influenza A or B with standardized reagents and test kits supplied by the WHO in cooperation with the CDC. Collaborating laboratories then forward influenza samples with accompanying epidemiologic information to one of four international collaborating centers for reference and research (located in Australia, the United Kingdom, Japan, and the United States).

A major goal of the WHO network is to monitor the emergence and spread of variant viruses using a defined set of phenotyping and genotyping assays. Such assays enable the four collaborating centers to (1) screen influenza samples in a short period of time, (2) determine if significant influenza activity is associated with the isolation and spread of variant viruses, and (3) judge whether there are reduced postvaccine immune responses to variant viruses in individuals who have received the current vaccine. Cumulative data from the relevant geographic locations are then used by the WHO to make vaccine strain recommendations for the Northern and Southern Hemispheres on a semiannual basis.

Although gaps exist in the global surveillance of influenza, due to sparse sampling in underdeveloped countries and lack of sampling in animal reservoirs, currently there is no shortage of samples collected by the WHO program. On a worldwide scale, more samples are collected each year (~170,000) than are characterized (6,500) by the four collaborating centers. What is limiting is the laboratory work force for performing repetitive tasks on the available samples and then inputting the results into databases. It is important to realize that the testing procedures used by the WHO are reproducible among independent laboratories, which avails them to standardization, scale-up, and automation.

Thus, what influenza surveillance and research efforts really need are high-throughput automated laboratories and database resources that offer seamless integration of "digital" and "physical" tasks from start to finish. This includes tasks pertaining to (1) collecting samples and recording epidemiologic observations in the field, (2) screening samples to see whether they contain influenza A or B, (3) growing and titering viral samples, (4) phenotyping viral samples using a number of key assays, (5) genotyping RNA segments in viral samples, (6) archiving viral samples, (7) storing observations and laboratory results in relational databases, (8) analyzing data with informatics tools, and (9) sharing key findings with scientific collaborators, public health officials, and vaccine manufacturers on a timely basis. An automated influenza laboratory and database effort may lead to more accurate influenza vaccines in the short term (with more yearly information) and broader flu vaccines in the long run (with more cumulative information).

ENSURING SAFE FOOD

Recent surveys conducted by the CDC show that foodborne infectious diseases account for 76 million illnesses, 325,000 hospitalizations, and 5,000 deaths in the United States each year. Three known pathogens—Salmonella, Listeria, and Toxoplasma—cause 1,500 deaths each year while unknown infectious agents account for 3,200 deaths (Mead et al., 1999). Most foodborne outbreaks in the United States have been confined to relatively small numbers of people, mainly because of early detection and aggressive intervention in the course of events. Despite such luck, however, certain trends pose new and mounting threats to the food supply such as (1) the consolidation of food processors, (2) the increasing import of all types of foods, (3) the consumption of more meals away from home, (4) the emergence of new strains of plant and animal pathogens, (5) the shift from multicrop to single-crop farming, (6) the threat of bioterrorism, and (7) the inevitable rise of global commerce.

The U.S. system for ensuring safe food is composed of two complementary arms. The enforcement arm deals with rules and regulations for manufacturing, transporting, storing, importing, inspecting, and testing foods. It is administered by a patchwork of over a dozen federal agencies, often with limited resources and antiquated procedures. The investigatory arm deals with managing acute outbreaks and formulating recommendations for minimizing future ones. Federal agencies, such as the CDC and the Food and Drug Administration, have the authority to initiate investigations of outbreaks but can do so only if state and local public health agencies have adequate systems in place for detecting and reporting them. Therefore, the CDC offers informatics resources (such as Internet-based software and computerized databases) to state and local public health agencies in order to facilitate the reporting of outbreaks.

At the federal level, new methods from molecular epidemiology are now offering rapid and accurate tools for investigations of outbreaks and follow-up surveillance. The CDC and participating public health laboratories can now subtype enteric pathogens and then deposit their molecular fingerprints into a computerized database, enabling queries from other laboratories on related outbreaks and pathogenic strains. With such tools the public health infrastructure is able to compile up-to-date records of Escherichia coli O157:H7 outbreaks, for example. In addition, the centralized system helps to organize data on outbreaks that are dispersed over large geographic areas, involve many small clusters of cases, and/or occur in places with limited investigatory resources.

Most foodborne outbreak investigations use combinations of epidemiologic, laboratory, and informatics tools at every step. The customary steps include (1) identifying the circumstances that triggered the investigation, (2) ascertaining the outbreak's size, (3) examining the individual cases, (4) identifying the common links and risk-related causes, (5) implementing measures to control the outbreak, and (5) formulating recommendations for future prevention. Outbreak investigations can be difficult because of the increasing number of new pathogens and risk factors and also because major outbreaks crossing state and national boundaries involve overlapping enforcement and investigatory jurisdictions.

Ensuring a safer food supply will necessarily entail increased testing of various foods plus rapid investigation and control measures once outbreaks have occurred. The colloquium therefore considered scientific and risk assessment approaches for protecting the food supply and the applications of high-throughput automated laboratory and informatics resources in reaching such goals.

MITIGATING BIOTERRORISM AND BIOWARFARE

The problems of bioterrorism and biowarfare are particularly demanding because their solution involves many scientific disciplines plus institutional coordination on federal, state, and local levels. From medical, public health, and emergency response perspectives, bioweapons may unleash viruses, bacteria, fungi, and/or toxins that cause unusual symptoms and highly virulent forms of disease. To save lives, the first responders to an attack will require accurate and rapid information for diagnosis, treatment, and quarantine measures if indicated. From law enforcement and national security perspectives, threats may come from rogue states and/or terrorist groups that crisscross the jurisdictions of several government agencies. And because each agency has its own operating procedures, the application of technologies across federal, state, and local jurisdictions will require careful planning. From political and economic perspectives, biological arsenals are classified as weapons of mass destruction and they can be stockpiled at a fraction of the cost of chemical and nuclear weapons. In response to this, international laws that ban the proliferation of bioweapons will require the means for inspection and verification. It will thus be incumbent on policymakers and scientists to define laboratory technologies that do not infringe on commercial enterprises (e.g., pharmaceutical manufacturers) yet remain fully capable of spotting significant violations.

The dismantling of the former Soviet Union and the system of two opposing superpowers has led to an uncertain world order. It now includes one global superpower and numerous responsible governments—yet it also includes several rogue states, multiple religious fringe groups, and a few shadowy international syndicates that are forming new networks and posing new challenges to national and global security. Today, at least 17 countries are known to be developing or producing bioweapons, and the list may be expanding (Alibek, 1999). The former Soviet Union was engaged in a secret and extensive offensive biological weapons program directed against personnel, materiel, animals, and plants and involving many organizations and facilities in Russia and the republics. Tens of thousands of personnel were involved—in the scientific academies; government departments responsible for defense, agriculture, and public health; and in an entity known as Biopreparat, which alone had over 32,000 employees. Presently, it is not clear whether all elements of the program have been halted, especially in the Russian Ministry of Defense. Given the severe economic conditions since the collapse of the former Soviet Union, there are also significant proliferation risks for trained personnel and weapons materials through recruitment of former weapons scientists by rogue states. In contrast, the United States ended its bioweapons program in 1969 and now retains few living experts with first-hand knowledge of such programs. Such loss of expertise makes it even harder to deal with various evolving threats.

To evaluate the threats of bioterrorism and biowarfare, one must understand basic steps in the weaponization process. Exact details of each step will depend on the scale of the effort, but in general they include (1) choosing infectious agents, (2) isolating useful seed stocks and pathogenic biotypes, (3) formulating stabilizers and scatter enhancers, (4) growing and packaging enhanced bioagents, (5) storing and monitoring bioweapons, (6) maintaining one or more reliable delivery systems, and (7) identifying targets and contingencies. The detailed process of building fully strategic bioweapons may take years, yet simpler ones may be produced in less time. Since weather conditions influence the spread and viability of infectious agents, sophisticated bioweapons may contain pathogens with different physical stabilities and methods of dispersal.

In the United States as few as 10 to 100 afflicted people may stretch hospital, public health, emergency response, law enforcement, and national guard services beyond their current limits. Even larger attacks involving major metropolitan areas may literally require the delivery of tons of antibiotics to exposed persons within days, totally overwhelming current response capabilities. In the twenty-first century, mitigating the threats of bioterrorism will thus demand sizeable laboratory and informatics resources, which can be organized in terms of four overall phases.

First, in preventing an attack, responsible governments may rely on the ability to fingerprint infectious agents efficiently with high-throughput automated laboratories. An extensive database of molecular fingerprints would give responsible governments a new means of rapid attribution and therefore deterrence. It would also put rogue states, religious fringe groups, and international syndicates on notice that there is little chance to evade blame for any bioattack.

Second, in the unfortunate event of an attack, U.S. public health laboratories may be overwhelmed within the first few moments—quite simply because there would be too many samples to process and test within hours. With manual laboratories it would be impossible to answer even the simplest of questions. How many different infectious agents were released? How do they differ? What are the best initial ways to treat those afflicted? Information from high-throughput automated laboratories may reduce confusion and save lives by offering timely testing in acute situations.

Third, in the aftermath of an attack, U.S. public health, agricultural, and law enforcement officials would need accurate answers to yet another set of questions. What is the stability of each infectious agent? What are their geographic boundaries? What are the effects on animals and plants? What are the molecular fingerprints and origins of the agent(s)? Information from high-throughput automated laboratories may speed the recovery process by offering testing for cleanup and investigatory operations.

Fourth, in response to an attack, U.S. law enforcement officials may need to collect evidence in accordance with chain-of-custody procedures. Intelligence agencies and military services may need to make accurate attributions and then take swift actions to protect national security. Information from high-throughput automated laboratories and their associated databases may prevent further attacks by rapidly pinpointing sources and locations.

APPLICATIONS BEYOND INFECTIOUS DISEASES

Quite clearly there are scientific needs for high-throughput automated laboratories and informatics resources in other areas, such as (1) human genetics, (2) molecular medicine, and (3) pharmaceutical screening. The colloquium therefore considered the scientific needs in these areas by inviting participation from a widespread scientific and technical constituency. For instance, researchers are now considering how to approach problems in medicine and biology after the human genome is fully sequenced and 90 percent of its SNPs are found. Such data are expected to become available within the next few years from corporate and governmental sequencing programs, opening unprecedented opportunities for molecular medicine and pharmaceutical screening. It will therefore be essential to have high-throughput laboratory and informatics resources to take advantage of such achievements.

Data for most problems in medicine and biology can be classified in terms of three spaces. Epidemiologic space pertains to exposures, expressions, and outcomes in human populations over time—for example, the incidence of certain cancers that are due to environmental exposures, inherited traits, and/or acquired mutations. Phenotypic space pertains to physical properties or biomarkers at the cellular level—for example, the expression of abnormal proteins and molecular histologies in cancer cells. Genotypic space pertains to the actual DNA sequences, whether they represent normal variations within a population or abnormal ones—for example, the presence of oncogenes and translocations in chromosomes that increase cancer risks. Certain diseases without environmental factors that are due to one abnormal protein, or that are caused by one SNP may have the simplest spaces to explore and understand. Yet such simple spaces are believed to represent only the tip of the iceberg in medicine and biology. To explore a wider range of diseases and develop tailored medications, it will be necessary to discover correlations and mappings in more intricate spaces.

One approach is to study simple inherited diseases in relatively homogenous human populations and then extend the scientific approach to more heterogeneous populations. Another approach is to use living organisms that reproduce rapidly (e.g., bacteria, yeast, mice) to explore a small group of mutations and fitness, with rapid-design strategies known as DNA shuffling. The colloquium examined ways to build high-throughput laboratory systems for supporting such approaches. It also examined the available informatics tools for discovering relationships from enormous phase spaces that contain various epidemiologic, phenotypic, and genotypic data.

OVERVIEW OF TECHNOLOGIES AND APPROACHES

All the necessary building blocks are available for establishing petabit-generating user facilities in medicine and biology. In fact, for laboratory automation and robotics, key technologies have matured to the point where manufacturers and consumers alike are asserting that products must conform to a single interconnection standard recently adopted by the American Society for Testing and Materials (ASTM). By conforming, products from one manufacturer would be "plug and play" with products from another, simplifying high-throughput automated systems. Thus, the real challenge is to identify compelling problems and then build integrated systems that are flexible and capable of utilizing advances in technology as they become available.

The available technologies and approaches considered by the colloquium included (1) laboratory automation and robotics, (2) interconnection standards, (3) genomic sequencing, (4) flow cytometry, (5) microtechnologies, (6) advanced diagnostics, (7) mass customized testing, (8) databases, (9) informatics, (10) mathematical modeling, (11) risk assessment, and (12) molecular breeding. As summarized below, each has applications in fighting infectious diseases, ensuring safe food, mitigating bioterrorism and biowarfare, and facilitating work on human genetics and molecular medicine.

Laboratory Automation and Robotics

Since its founding in 1995, the Association for Laboratory Automation has held annual conferences on the rapidly advancing fields of clinical and laboratory automation. The recent LabAutomation'99 conference was attended by over 2,200 participants and 85 corporate exhibitors. In addition to scientific and technical presentations, the exhibition halls displayed a large variety of commercial automation hardware, such as robotic arms/conveyers, bar code readers, material supply modules, liquid handlers, plate washers, plate readers, incubators, filtration modules, centrifuge modules, genomic sequencers, flow cytometers, image analyzers, and disposal stations. Companies also displayed a number of ready-made systems that integrate hardware, software, and reagent streams for complete high-throughput procedures. A few examples of such offerings included (1) systems that grid filter papers and glass slides with arrays of molecular probes, (2) systems that identify, pick, and culture bacterial colonies, and (3) systems that screen compounds for drug activity and biological toxicity. The key point is that flexible hardware from such systems can serve as components in petabit-generating user facilities.

Interconnection Standards

Many companies now only sell proprietary software for communicating with their powerful hardware, yet no single company offers a full range of high-throughput components. This fragmentary environment has often led to frustrations because it takes an inordinate amount of time and effort to interconnect hardware from different manufacturers. To overcome this obstacle, the ASTM has adopted the Laboratory Equipment Control Interface Specification (LECIS). The new interconnection standard was formalized in 1999 (as ASTM E1989-98) and, already, several leading laboratory automation and pharmaceutical manufacturers have announced their intentions to support it. In the coming months, laboratory automation and robotics vendors will offer an increasing number of LECIS-compliant products, with many plug-and-play features like those expected in personal computers (Committee E01, 1999).

The LECIS instrument control concept is based on interactions between a single controller and any number or type of standard laboratory modules under its command. Essentially, LECIS organizes equipment behavior into a small number of "states" and also defines a small number of "messages" for managing transitions between these states. These two elements are fully independent of programming languages and physical scales (e.g., length, volume, time). It is therefore feasible to build flexible instruments with older macrotechnologies (e.g., pipettes, test tubes, 96-well plates); newer microtechnologies (e.g., microchannels, microchambers, microdetectors, microarrays); or modular combinations of both. Various hardware devices that comply with the ASTM standard would be plug and play with others using it, and, like today's personal computers, large instruments could house duplicate modules for rate-limiting procedures and expansion slots for adding new capabilities over time.

LECIS also enables an extension known as the Device Capability Dataset (DCD), which describes unique characteristics for a particular piece of equipment. To facilitate a plug-and-work environment, each DCD would contain information on the equipment's function, identification, physical characteristics, location, communication ports, commands, events, exceptions, errors, resources, and maintenance. Such information is also independent of programming languages and physical scales, permitting flexible integration of macro- and microtechnologies.

Genomic Sequencing

The colloquium considered two different technologies for high-throughput genomic sequencing. One is based on chain-terminating amplification and gel electrophoresis, which has been revolutionized by microcapillary structures. The other is based on selective hybridization to complementary DNA oligomers, which are formed and immobilized on microchip arrays. Both technologies are compatible with high-throughput automation, yet in many semiautomated laboratories the tasks of sample preparation and data analysis have limited their full potential.

Apt uses of chain-terminating amplification and gel electrophoresis are the sequencing of long segments of stable DNA (e.g., mammalian and bacterial genomes) and the sequencing of short segments of variable DNA/RNA (e.g., viral genomes). Commercial microcapillary gel-based instruments are capable of processing approximately 300,000 bases per day, and such outputs are increasing as the number of cycles per day increases. Current efforts to sequence human, animal, and plant genomes use several hundred instruments in parallel.

Apt uses of complementary DNA oligomers are the sequencing of short segments of variable DNA/RNA (e.g., viral genomes) and, quite importantly, the real-time monitoring of many different complementary DNA (cDNA) molecules in living cells. Commercial microchip-based instruments are capable of processing approximately 100,000 bases per day, and such outputs are increasing as the density of oligomers per microchip increases. Current efforts to monitor cDNA in Mycobacterium tuberculosis before and after exposure to antibiotics, to determine the frequency of SNPs in human populations, and to find genetic differences in normal versus cancer cells use a variety of microchip instruments.

The ability to perform various types of high-throughput genomic sequencing completely alters the scale of possibilities for infectious disease control and threat reduction. For example, an automated influenza laboratory could integrate and use commercial instruments to sequence

RNA segments for every sample tested each year. An automated food safety and/or bioterrorism mitigation laboratory could also integrate and use commercial instruments to fingerprint literally thousands of bacterial pathogens within 24 hours after an event.

Flow Cytometry

In essence, flow cytometry uses hydrodynamic focusing to guide a thin column of fluid through laser beams. When samples pass through the lasers, photons are scattered and emitted at various angles and intensities that reveal physical-chemical features about each sample. Because thousands to millions of samples can pass through the laser beams within seconds, flow cytometry utilizes signal averaging and thereby produces large signal-to-noise ratios. Several companies now offer flow cytometers that are designed to work in conjunction with laboratory automation and robotics. The colloquium therefore considered various roles for this flexible technology.

Detectable samples may be as small as individual molecules and viral particles or as large as intact bacteria and animal cells. In practice, biological samples are often labeled with nonspecific DNA intercalaters and/or specific antibodies that fluoresce under laser light, thereby providing quantitative information on DNA sizes and protein expression. Alternatively, biological samples may be mixed with special microspheres (e.g., 100 to 1,000 different kinds at once) that are painted with different fluorescent dyes (color multiplexing) and covered with different capture molecules on their surfaces (probe multiplexing). The mixtures are then flowed in order to see which microspheres captured the samples. Such methodologies are analogous to performing multiple enzyme-linked immunosorbent assays (ELISA) on the same sample at once, and they are now commercially available. Once samples flow through the laser beams, the fluidic column may be divided into microscopic droplets by piezoelectric or inkjet devices. This permits droplets with samples to be dispensed in accordance with detection events. Flow cytometry thus provides a fast means for sample detection, separation, and purification.

The ability to perform various types of high-throughput sample characterization completely alters the scale of possibilities for infectious disease control and threat reduction. For example, an automated influenza laboratory could integrate and use commercial instruments to phenotype viral samples (i.e., determine subtypes and/or immunologic relatedness) or genotype (i.e., select optimal polymerase chain reaction primers) within seconds. An automated food safety and/or bioterrorism mitigation laboratory could also integrate and use commercial instruments to phenotype and genotype literally thousands of bacterial pathogens within 24 hours after an event.

Microtechnologies

Forward-looking universities, national laboratories, and companies are fostering programs to model, design, test, and manufacture miniaturized devices that perform chemical and measurement-based operations. For medical and biological researchers, this push to miniaturization holds great promise. It is comparable to revolutions seen in the computer electronics industries, where vacuum tubes gave way to transistors, which then gave way to integrated circuits. Test tubes have already given way to microtiter plates with 96, 386, and 1,536 wells. Thus, the next and inevitable step is for microtiter plates to give way to miniaturized and integrated laboratories-on-a-chip (LOC), where performances may be rated by the number of chemical instructions per second.

With miniaturization comes the ability to integrate operations such as separation, reaction, detection, and signal processing and build them into micron- and submicron-sized areas. In general, heat and mass transport constraints on chemical reactions are insignificant from microliter (10^{-6}) to femtoliter (10^{-15}) scales, so LOC devices can perform subnanomolar-scale reactions and syntheses as well. The development of successful and flexible LOC devices will require the efforts from several disciplines, including physics for modeling microhydrodynamics and fluid mechanics; engineering for developing chemical circuits, sensitive detectors, control, and power systems; computer science for dealing with enormous quantities of data from microtechnologies; and biologists for scaling down assays and finding useful applications. The colloquium therefore considered various applications for microtechnologies, particularly with regard to the advanced diagnostics summarized below.

Advanced Diagnostics

In national security, public health, and medical practice areas, there are growing demands for rapid, one-step, throwaway devices that can diagnose infectious diseases on the spot. Ideally, these point-of-use devices would be capable of detecting infectious disease agents without having to grow them or amplify their genomes. They would also have detection limits equal to the agent's infectious dose 50 percent (ID_{50}) for humans and/or animals, enabling them to signal a bioterrorist attack as it occurs. Such diagnostic devices are now under development by several corporate and government-based programs, but many desirable capabilities are still futuristic. For instance, many biowarfare agents have ID_{50}s in the range of 10^1 to 10^3 organisms, and over the next decade it will remain difficult to build point-of-use devices with such minuscule sensitivities. Thus, for the foreseeable future, point-of-use devices may offer help only in diagnosing exposed and/or acutely ill people.

During acute influenza illnesses, people shed viruses in their respiratory secretions at titers ranging from 10^3 to 10^9 tissue culture infectious doses per milliliter. Such in vivo titers correspond to viral-associated protein concentrations of 10^{-15} to 10^{-9} moles per liter, which falls within the detection limits of cutting-edge technologies being developed by corporate and government-based programs. Such benchmarks are typical for many other bioterrorist agents, suggesting that advanced diagnostics may offer help against them as well. The colloquium therefore considered how advanced diagnostic devices may function in conjunction with high-throughput automated laboratories and their associated databases.

One concept is to mass produce rapid, one-step, throwaway devices as combined screening dipsticks and sample containers. For the global influenza surveillance network, for example, such devices would be used to test for influenza A and B in humans or animals on the spot. They would be the size of small matchboxes, display bar codes for identification, provide simple (+/–) answers, and contain testing and sample holding chambers. For positive samples one would record key epidemiologic information and forward the device to the automated influenza laboratory for subsequent characterization. For negative samples the devices would be discarded. Such procedures would save time and expense by reducing work on negative samples.

Another concept is to use advanced diagnostic devices for rapidly screening and diagnosing people during the acute phases of bioterrorist attacks. For positive individuals, emergency medical personnel could initiate antibiotic therapies on the spot, which may help save their lives. The diagnostic/sampling devices would then be sent to an automated forensics laboratory, where molecular fingerprints and antibiotic sensitivities would be used for characterizing the overall attack and saving more lives.

Mass Customized Testing via the Internet

Because of Internet connectivity and the availability of worldwide commerical shipping services, the first high-throughput automated laboratory may be situated in the United States yet be accessible from practically any geographic location. Use of the automated laboratory would begin by downloading a set of process control tools (PCTs) over the Internet and installing them on personal computers. These software tools would be written in a platform-independent language (e.g., Java), would be run in conjunction with web browsers, and could create a flexible environment where the automated laboratory functions like an army of programmable technicians (i.e., mass customized testing). Since collection sites and laboratory facilities may be located on entirely different continents, the system would necessarily operate on a nonreal-time rather than a real-time basis. The key point is that nonreal-time systems demand small communication bandwidths that are supported by today's Internet communications protocols, whereas real-time systems demand large bandwidths that are redundant, fail-safe, and not widely available.

In the nonreal-time environment, users would collect samples and then use convenient computerized tools for recording various epidemiologic data and programming laboratory procedures. Testing instructions and background information would arrive over the Internet, and barcoded samples would arrive via airfreight or other convenient means. Within just a few days the tests would be set up and performed by high-throughput automation in accordance with the "assay scripts" that arrived over the Internet. Results would then be deposited in the database resource and made available to users and others with access privileges. All testing and informatics services would arise from three levels of control (high, intermediate, and low) and would take advantage of the interconnection standards mentioned above.

High-level controls for the automated facility would act like those found in interactive websites, with point-and-click objects and pull-down menus. Outside the facility, users would click on PCT icons to implement and customize various stepwise tasks. Inside the facility, supervisors would click on other PCT icons to manage workloads and maintain automated machinery in optimal working order. A complete set of PCTs, for example, pertaining to access, operation, documentation, submission, storage, analysis, privileges, and accounting, would offer seamless integration of digital and physical tasks from start to finish (Layne and Beugelsdijk, 1998).

Intermediate-level controls would reside in laboratory-based computers, whose functions remain virtually transparent to outside users. To schedule daily "assay runs," laboratory supervisors would use tools from operations research to simulate automated instrument activities and various constraints (i.e., timing, capacities) imposed by testing procedures. Basically, assay scripts would be modeled by linear equations, fit into a large matrix with others, and solved by numerical algorithms. Typical assay runs may include as many as 10,000 precisely timed and ordered tasks, which far exceeds human capacities for effecting work. Optimized schedules would then go to an instrument's controller that synchronizes and governs the intricate flow of samples, reagents, and supplies.

Low-level controls would also reside in the facility's robotics and automation modules and serve two purposes. The first is to drive internal components like actuators, detectors, and servomotors and coordinate their internal electromechanical activities. The second is to communicate with an instrument's controller and follow its commands, as set forth in the interconnection standard mentioned above.

From the perspective of ease and maintenance, it is often better to build several different high-throughput automated systems that work together instead of one larger one. For this reason the automated influenza, food safety, and threat reduction laboratories would likely use a series integrated systems that work as a seamless unit while enabling mass customized testing via the Internet. Depending on applications, the various systems would focus on, for example, growing, phenotyping, genotyping, biotyping, drug sensitivity testing, and molecular fingerprinting. Such systems may utilize various combinations of immobile laboratory automation and robotics technologies, as well as mobile microtechnologies and advanced diagnostic devices. Since the associated databases would also utilize the Internet to receive and send information, they may be located at any geographic site.

Databases

New databases for high-throughput automated laboratories must be set up to store and organize a wide variety of structured data. For example, epidemiologic data may include clinical observations on humans or animals, references to geographic locations and times, and notations on various samples being collected. Such data may come in the form of ASCII text, standardized questionnaires, graphical representations, and audio/video records. Phenotypic and genotypic data may include results from various automated assays and quality controls, programmed instructions for performing such assays and controls, and notations on special reagents or procedures. Such data may come in the form of tabular, numerical, and image-based records. Fortunately, programming languages have reached the point where it is feasible to set up powerful databases that work in conjunction with the Internet. The colloquium therefore considered various technological and proprietary issues pertaining to such undertakings.

High-throughput automated laboratories will require databases that are flexible, scalable, and secure. These requirements are fulfilled to various degrees by commercial database systems that are written in object-oriented or relational languages. Object-oriented systems (e.g., Object Database Management Group) are good at expressing elaborate relationships among objects and manipulating data but are less suited for storing enormous quantities of data. On the other hand, relational systems (e.g.,

Symbolic Query Language) are good at storing and retrieving enormous quantities of data but are less suited for handling elaborate relationships and manipulations. Security issues deal with averting the loss and corruption of data and preventing the unauthorized use of data, whether inadvertent or malicious. Both object-oriented and relational database systems have extensions that enable digital certificates, secure sockets, and secure partitions.

Commercial database systems further support the Extensible Markup Language (XML). This documentation standard is ideal for object-oriented and relational databases that must handle a wide variety of structured data. With XML all structured data become self-describing, platform independent, and transformable into any format. The XML standard also permits multiple pointers, links, and references to multiple sources of data, equipping it to handle many of the problems considered by the colloquium (World Wide Web Consortium, 2000).

With automated laboratories changing the means by which research is conducted, it will be important to maintain traditional rewards for users. At the heart of this system is the freedom to decide how to share data and new information, which can lead to scientific publications and credit for discoveries involving intellectual property. For each category outlined below, data ownership and privileges may be assigned according to the source of financial support.

For the closed category, data would belong solely to the commercial organization (e.g., a pharmaceutical company) that submitted samples and assay scripts and paid for research or testing services. Upon completing such work, the automated laboratory would encrypt and forward all of the raw data to the purchasing organization. Afterwards, it would be the organization's responsibility to manage the security of its private property. For a period of time the automated laboratory would also maintain a secure copy of the digital records to assure redundancy and integrity in accordance with contractual agreements.

For the principal investigator category, data would belong to the person receiving government grant support for a reasonable period of time, say for as long as 2 to 3 years after the grant ends. Good digital practices would be tied to ongoing grant support, requiring each investigator to maintain his or her database records in an orderly manner. After the time embargo had expired, relational links would be attached to the investigator's digital records and the information would become available to others.

For the consortia category, data would belong to all of the collaborating investigators for a reasonable period of time, as suggested above. The collaborators also would have responsibility for maintaining their digital records in an orderly manner, most likely under the supervision of the group's database manager. After the time embargo had expired, the organized information would become available to others as well.

For the open category, data and its associated links would belong to the public after assuring its quality. The digital records would come from voluntary submissions and time-embargoed data that would be released automatically. The main issue would be maintaining backup copies to assure integrity, plus deciding how to inventory the data and build relational links.

Informatics

Mining enormous quantities of data from high-throughput automated laboratories (i.e., petabits) poses an open-ended challenge. It requires that users have fast computers and Internet links for undertaking their work. It also requires that users have just the right software algorithms for analyzing their data. Today, researchers have affordable personal computers that perform 10^9 floating point operations per second (gigaflops), and for large jobs they can rent time on expensive supercomputers that perform 10^{12} floating point operations per second (teraflops). They have connections to Internet backbones that transmit 10^9 bits of information per second (gigahertz) and local fiber optic networks that approach such bandwidths. In addition, a growing number of informatics companies are offering new types of software for mining medical and biological data. Just a few years ago these programs were affordable only by global pharmaceutical companies, but prices have fallen to the point now where university-based researchers can buy them. Nevertheless, many informatics tools still need improving, some types are missing, and certain types await definition. The colloquium took a problem-oriented approach to the development of such resources.

In utilizing high-throughput automated laboratories it will be valuable to have two types of informatics tools. One type will be designed to work with smaller datasets, which are generated on a day-to-day basis, whereas the other will work with larger datasets that amass over longer periods of time. This division is practical because short-term informatics tasks are different than long-term ones. Short-term tasks pertain to checking experimental assays against quality controls, spotting obvious patterns or biases, and programming new assays as necessary. Long-term tasks pertain to discovering correlations and order in multidimensional spaces that include epidemiologic, phenotypic, and genotypic data.

Understanding how molecular sequences at the genetic level determine such things as protein structures, enzymatic activities, and antigenic identities is one of the grand challenge problems in medicine and biology. It is also a problem at the crossroads of infectious diseases, threat reduction, and molecular medicine. For example, researchers have attempted to look for long-term trends in successful mutations for rapidly drifting and shifting viruses such as influenza. Their ultimate goal is to forecast "hot strains" that will cause major epidemics or explosive pandemics and thereby expedite the production of new vaccines against such strains. At the present time, however, researchers have an insufficient database for making solid forecasts on influenza's next step. With sparse data linking the genetic, structural, enzymatic, and antigenic properties of influenza, they have no means for predicting successful mutations. Still needed are many more records containing complete epidemiologic, phenotypic, and genotypic information on influenza isolates over several years. Also needed are new informatics tools and algorithms that can correlate and display enormous quantities of genetic, structural, enzymatic, and antigenic data in clear ways.

Leveraging important scientific problems for the development of informatics tools makes a great deal of sense because fundamental questions in medicine and biology are interconnected at certain levels. For example, influenza has a small RNA-based genome (i.e., kilobases) with many variations, whereas humans have a large DNA-based genome (i.e., gigabases) with comparatively few variations. Nevertheless, the informatics problems associated with finding order by correlating enormous quantities of epidemiologic, phenotypic, and genotypic information are basically the same for the two entities. If properly designed and implemented, informatics tools for influenza will thus carry over to other important problems in infectious disease and molecular medicine.

Mathematical Modeling

Important problems in medicine and biology often exhibit nonlinearity, complexity, variability, and noise. As already discussed, one aspect of attacking such problems is to build high-throughput automated laboratories that generate enormous quantities of data. Another aspect, however, is to formulate mathematical models that extend intuition about the problem and further organize the collection, generation, and interpretation of data. The colloquium considered various roles for mathematical models in infectious diseases and epidemiology, with the understanding that underlying elements apply to food safety, bioterrorism mitigation, and molecular medicine.

The first role is to understand the complex nonlinear behavior of the problem and determine how certain variables and/or parameters influence the spread and control of infectious diseases. In this regard, epidemiologic models are built on simple assumptions that nevertheless simulate complex problems. They can therefore be used to build intuition and insight before starting high-throughput research efforts.

The second role is to guide data collection efforts and thereby create optimal inventories of data. Different infectious disease and epidemiologic models can be used to determine the relative importance of each parameter and its sensitivity to change. They can then be used to prioritize data collection efforts during high-throughput research efforts.

The third is to organize and map enormous phase spaces that are composed of epidemiologic, phenotypic, and genotypic data. One method is to discover relationships by statistical inference. Another is to develop models that may account for the data and then view the data from such perspectives. Infectious disease and epidemiologic models can therefore be used to look for unifying patterns or principles from high-throughput research efforts.

The fourth is to make forecasts or develop interventions with models that are validated against epidemiologic, phenotypic, and genotypic data. With sufficient information, for example, it may become feasible to forecast the next influenza pandemic or to develop broader vaccines with efficacy against many different influenza strains. Infectious disease and epidemiologic models can therefore be used to mitigate risks and save lives.

Risk Assessment

The total volume and variety of the food supply in affluent countries like the United States are huge. It is therefore impossible and impractical to test every lot or type of food product—even with the help of many high-throughput automated laboratories. With this perspective the colloquium considered the key roles of risk assessment in utilizing expanded laboratory testing capabilities.

Traditionally, risk assessment has been used as the basis for decision-making in chemical safety and only recently has been considered useful in assessing exposures to microbiological pathogens and/or toxins in foods. The underlying assumptions are that (1) all agents are hazardous at some dose, (2) certain agents are hazardous at all doses, and (3) hazard magnitude depends on dose size. "Risk" is therefore defined as the probability that an agent's hazardous properties will be expressed at specified doses, yet such assessments are often plagued by uncertainties in the available data.

Risk assessment aims to determine the likelihood that a hazard will be expressed under specified conditions. For example, given that certain quantities of bacteria are found in lots of ground beef, there is a likelihood that a foodborne outbreak will occur. Risk assessment can be used to model the food supply and identify the products (or product sources) that pose the greatest risks to public health. Products posing the greatest risks would then be routinely sampled and tested by high-throughput automated laboratories and informatics resources. It is hoped that ongoing validation and adjustment of risk assessments over time would result in safer food supplies.

Molecular Breeding

Life processes consist of reproduction, mutation, and selection. One way of accelerating their exploration is by a technology known as DNA shuffling. Basically, one or more genes are cleaved into fragments and randomly recombined to create many novel genotypes. These sequences are then selected for one or more desired expressions or phenotypes. Subsequently, the selection process is used to identify genes that will become starting points for the next cycle of recombination. The process is repeated until genes expressing the desired properties and/or stabilities are identified.

In practice, DNA shuffling uses living organisms that reproduce rapidly (e.g., bacteria, yeast, mice) in order to explore a small group of mutations and fitness. The colloquium examined ways to build high-throughput laboratory systems for supporting such approaches. It also considered how informatics tools could be used for rapidly discovering relationships from phase spaces that contain various phenotypic and genotypic data. In many areas of biology and medicine, knowledge has been gained by studying a large number of natural mutations and their effects. In DNA shuffling, however, knowledge is now being gained by studying a relatively small number of clever mutations and their effects.

REFERENCES

Alibek, K. 1999. Biohazard. New York: Random House.

Committee E01. 1999. E1989-98 Standard Specification for Laboratory Equipment Control Interface (LECIS). West Conshohocken. Pa.: American Society for Testing and Materials (http://www.astm.org).

Layne, S. P., and T. J. Beugelsdijk. 1998. Laboratory firepower for infectious disease research. Nature Biotechnology, 16:825-829.

Mead, P. S., L. Slutsker, V. Dietz, L. F. McCaig, J. S. Bresee, C. Shapiro, P. M. Griffin, and R. V. Tauxe. 1999. Food-related illness and death in the United States. Emerging Infectious Diseases, 5:607-625.

Pablos-Mendez, A., M. C. Raviglione, A. Laszlo, N. Binkin, H. L. Rieder, F. Bustreo, D. L. Cohn, C. S. B. Lambregts-van Weezenbeek, S. J. Kim, P. Chaulet, and P. Nunn. 1998. Global surveillance for antituberculosis drug resistance, 1994-1997. New England Journal of Medicine, 338:1641-1649.

World Wide Web Consortium. 2000. XML 1.0 Second Edition Working Draft Released (http://www.w3.org).

PART I

INFECTIOUS DISEASES

3

The Application of Mathematical Models in Infectious Disease Research

Roy M. Anderson

The European countries have concerns similar to those of the United States regarding emerging or introduced infectious diseases in at least three specific areas. First, because London and other major cities in Europe serve as hubs for international travel, there is a continual risk of the introduction of new pathogens or new strains of endemic pathogens from other regions of the world. Second, the general public has become increasingly sensitized to the issue of food safety because of outbreaks of highly pathogenic strains of *E. coli*, and the prion-related "Mad Cow disease," with its associated disease in humans, new variant Creutzfeldt-Jakob Disease. Third, European governments are increasingly aware of a general vulnerability to acts of bioterrorism involving the deliberate release of dangerous pathogens.

This paper addresses the use of mathematical methods in research on infectious disease problems in terms of their emergence, spread, and control. Mathematical analysis is a scientific approach that can provide precision to complicated fields such as biology, where there are numerous variables and many nonlinear relationships that complicate the interpretation of observed pattern and the prediction of future events. Modern computational power provides us with extraordinary opportunities to sort through the growing volume of data being generated in the fields of biology and medicine. Many areas are ripe for the application of mathematical approaches such as the analysis of host and pathogen genome sequence information, the translation of genome sequence information into three-dimensional protein structure, the impact of global warming on ecological community structure, and analysis of the spread and control of infectious diseases in human communities. In the field of bioterrorism, appropriate methods can provide many insights into vulnerabilities and how to reduce these plus plan for the impact of a deliberate release of an infectious agent.

History is instructive. The first application of mathematics to the study of infectious diseases is accredited to the probabilist Daniel Bernoulli. Bernoulli used a simple algebraic formulation to assess the degree to which variolation against smallpox would change mortality in a population subject to an epidemic of the viral disease. The underpinning algebra, converted into modern terminology, is as applicable today in the context of assessing how different control interventions will impact morbidity and mortality as it was in Bernoulli's day. It provides a powerful tool to demonstrate or assess the use or impact of a particular intervention before it is put into practice.

POPULATION GROWTH AND TRAVEL AS FACTORS IN DISEASE SPREAD

Infectious diseases remain the largest single cause of morbidity and premature mortality in the world today. There are dramatic differences in the age distributions of populations throughout the world, and by and large this difference is dominated by mortality induced by infectious agents. The most dramatic example of this at present is the unfolding impact of AIDS in sub-Saharan Africa, where life expectancy in the worst-afflicted countries such as South Africa, Botswana, and Zimbabwe is predicted to fall to around 35 years of age by 2020 (Schwartländer et al., 2000). The evolution and spread of infectious agents are greatly influenced by the population size or density of the host species. In a historical context it is important to note the extraordinary opportunities today for the spread of infectious agents, given the rapid growth of the world's population over the past millennium (see Figure 3.1). Today, the major population growth centers are predominantly in Asia. In the recent past they have been in China, but India is projected to have the largest population in the near future. In the context of the evolution of new infectious agents and their propensity for rapid spread in growing communities, Asia is one location where we need to have high-quality surveillance for clusters in space and time of unusually high rates of morbidity and mortality that may be attributable to the emergence of a new pathogen or new strain of an existing infectious agent.

There are many other aspects to population structure and growth that are of relevance to the spread of infectious agents. One of these is where mega-cities will be located in the future, which again is Asia (see Table 3.1). These cities have very particular characteristics—small cores in the center

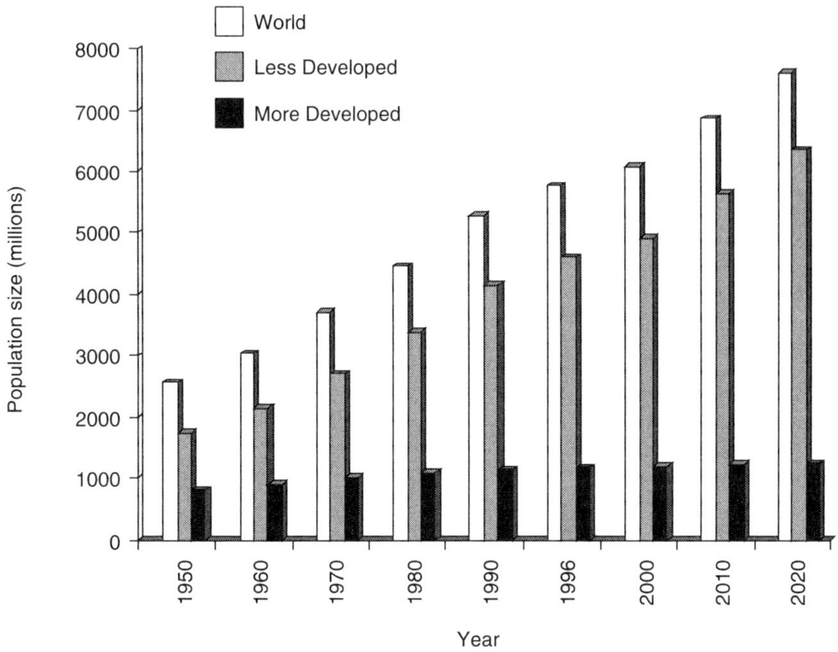

FIGURE 3.1 Projected world population size, 1950-2020.

with a degree of wealth and good health care in close proximity to very large surrounding peri-urban populations of extreme poverty, dense overcrowding, and poor sanitation and hygiene. These latter areas are important sites for the evolution of new infectious agents.

TABLE 3.1 Worldwide Increase in Megacities (number in different regions)

	1970	1994	2000	2015
Less Developed Regions				
Africa	0	2	2	3
Asia	2	10	12	19
Latin America	3	3	4	5
More Developed Regions				
Europe	2	2	2	2
Japan	2	2	2	2
North America	2	2	2	2

The rapid growth of international travel is also of great importance to the spread infectious disease. Of particular importance is travel by air, where transportation of infected individuals to the farthest corners of the globe can occur well within the time frame of the incubation period of virtually all infectious agents. The number of passengers carried by international airlines is growing linearly (see Figure 3.2). Closer examination of travel patterns show that the nodes and hubs of air travel are concentrated in particular centers, and cities in Asia increasingly serve as hubs for much international travel today.

EVOLUTION OF INFECTIOUS AGENTS

Quantitative study of the evolution of infectious disease is not well developed at present. Although in historical terms population genetics is a highly developed mathematical subject, the application of mathematical tools to the study of the evolution of, for example, influenza or antibody resistance is a neglected field with much research needed to provide predictive methods to help in policy formulation.

Evolutionary changes in infectious agents can occur rapidly under an intense selective pressure such as widespread drug therapy due to their short generation times relative to the human host. For example, a beta-lactamase-producing bacteria that is resistant to antibiotic treatment can,

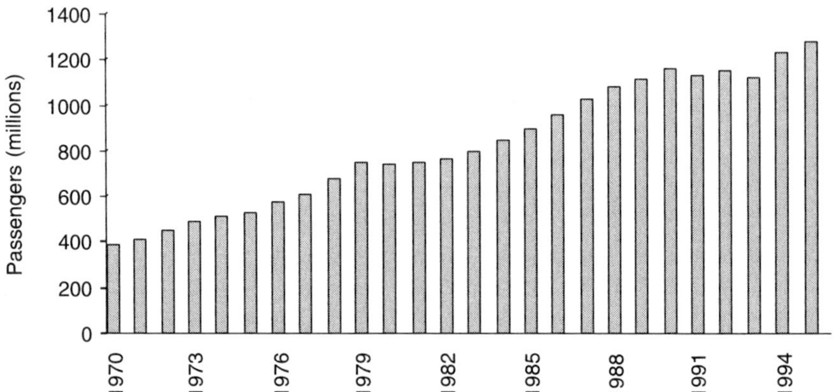

FIGURE 3.2 Development of worldwide scheduled air traffic, 1970 to 1995 (predicted annual growth annual growth of 5.7 percent from 1997 to 2001 with 1.8 billion passengers in 2001.

in a period of a few years, rise from a very low prevalence to a frequency of more than 80 percent. An example is given in Figure 3.3 for the spread of resistance in bacteria infecting children in Finland. It is difficult to understand the precise pattern of the emergence of drug resistance without correlating changes in frequency with the intensity of the selective pressure. That intensity is related to the volume of drug use in a defined community, the data for which are not always publicly available. Two time series—drug volume consumption and the frequency of resistance—are essential to interpret these patterns. Once this relationship is determined, predictions can be made, with the help of simple mathematical models, of how reducing the volume of drug use will affect the frequency of resistance. More generally, in the study of the evolution of infectious agents it is of great importance to quantify the intensity of the selective pressure concomitant with recording changes in the genetic constitution of the pathogen population.

PLOTTING EPIDEMIC CURVES

The plague killed one-third of Europe's population in the seventeenth century epidemic, and at that time it was one of the most pathogenic

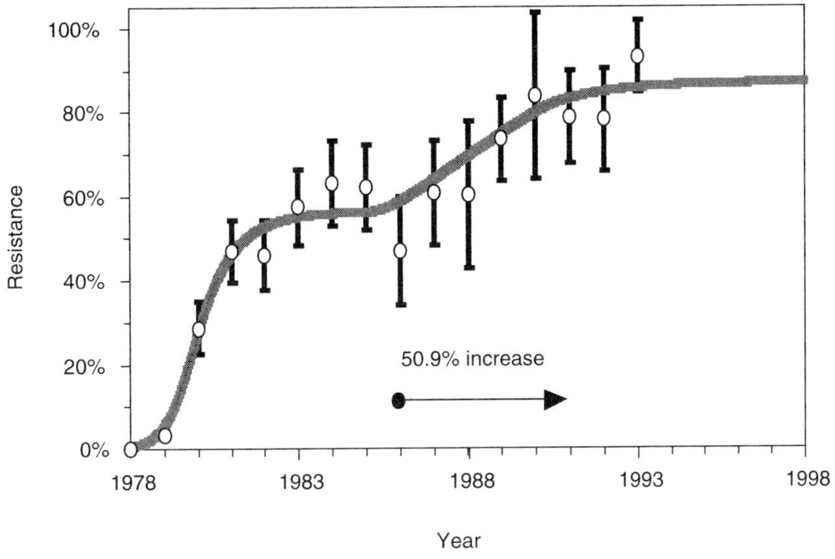

FIGURE 3.3 Frequency of TEMβ-lactamase-producing isolates of *Morexella catarrhalis* in Finnish children. Source: Nissinen et al. (1995).

organisms ever experienced by humankind. At the end of the last millennium an agent emerged that is both more pathogenic and has a greater potential to spread. The pathogen is HIV, which has a higher case fatality rate of nearly 100 percent, which is to be compared with a figure of roughly 30 percent for the plague. Mathematical models, specifically those describing epidemic curves, can serve as useful predictive tools to study the time course and potential magnitude of epidemics such as that of HIV.

An epidemic of an agent with a long incubation period, such as that found in HIV-1 infection and the associated disease AIDS, can be analyzed by constructing a set of two differential equations. One of the equations represents changes over time in the variable X denoting the susceptible population, and the other presents changes in Y denoting the infected population. The pair of equations can be solved by analytical or numerical methods, depending on the nonlinearities in their structures, to produce graphs that show changes in prevalence and incidence of infection over time. An example is plotted in Figure 3.4, which shows a simple bell-shaped epidemic curve that settles to an endemic state and the associated change in the prevalence of infection, which rises in a sigmoid fashion to a semistable state. The slight decline from the endemic prevalence is caused by AIDS-induced mortality.

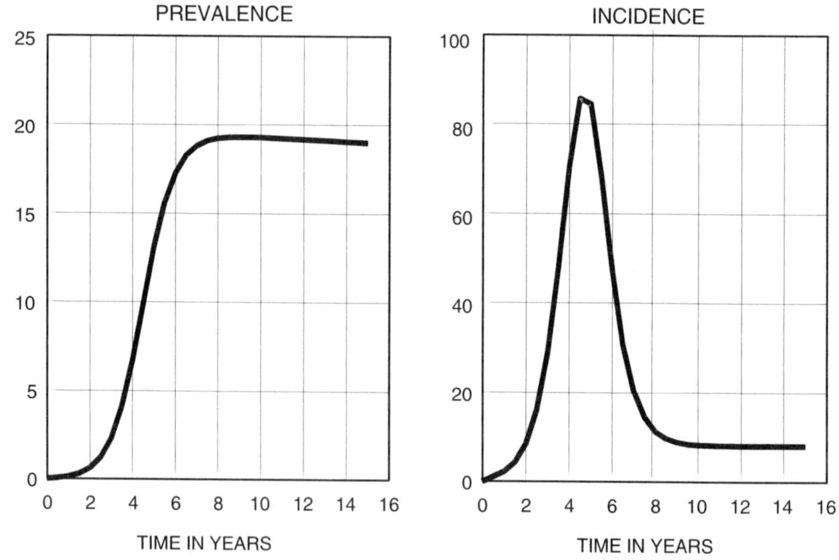

FIGURE 3-4 Relationship between prevalence and incidence in an HIV-1 epidemic.

Observed patterns are very similar to those predicted by this simple two-equation model. A specific example is provided by the time course of the HIV-1 epidemic in intravenous drug users in New York City. Simple epidemic theory suggests that saturation occurs in the course of the epidemic (as defined by the incidence of new infections), such that a natural decline occurs after an initial period of rapid spread due to a limitation in the number of susceptible individuals who are capable of contracting infection.

Concomitantly, a rise in the prevalence of an infection that persists in the human host will be sigmoid in form. In the absence of this knowledge of epidemic pattern, the decline in incidence could be falsely interpreted as the impact of interventions to control spread. This is exactly what happened in many published accounts of changes in the incidence of HIV-1 in different communities. Epidemiologists often interpreted a decline in incidence as evidence of intervention success. That may or may not be. The key issue in interpretation is to dissect the natural dynamic of the epidemic from the impact of any intervention. This example is a very simple illustration of the power of model construction and analysis in helping to interpret the pattern and course of an epidemic.

The principal factor determining the rate of spread of an infectious agent is the basic reproductive ratio, R_0. It is defined as the average number of secondary cases generated by one primary case in a susceptible population. The components of R_0 specify the parameters that control the typical duration of infection in the host and those that determine transmission between hosts (Anderson and May, 1991). An illustration of how R_0 determines chains of transmission in a host population is illustrated diagrammatically in Figure 3.5.

If this chain of transmission events is an expanding one, the quantity R_0 is on average greater than unity in value and an epidemic occurs. If $R_0 < 1$, the chain stutters to extinction. The quantity R_0 can be expressed in terms of a few easily measurable demographic parameters, such as the life expectancy of the host, the average age at which people are infected, and the average duration of protection provided by maternal antibodies.

Once such measurements are made for a given infection in a defined community, it is possible to derive estimates of the degree of control intervention required to suppress R_0 below unity in value and hence eradicate infection and associated disease. In the case of vaccine-preventable infections such as measles and pertussis, such calculations enable estimates to be made of the critical vaccine coverage required to block transmission and the optimal age at which to immunize. For example, the simplest and most accurate way to estimate the average age at infection prior to control is via a cross-sectional serological profile, which records the decay in maternally derived antibodies and the rise in immunity due to infection.

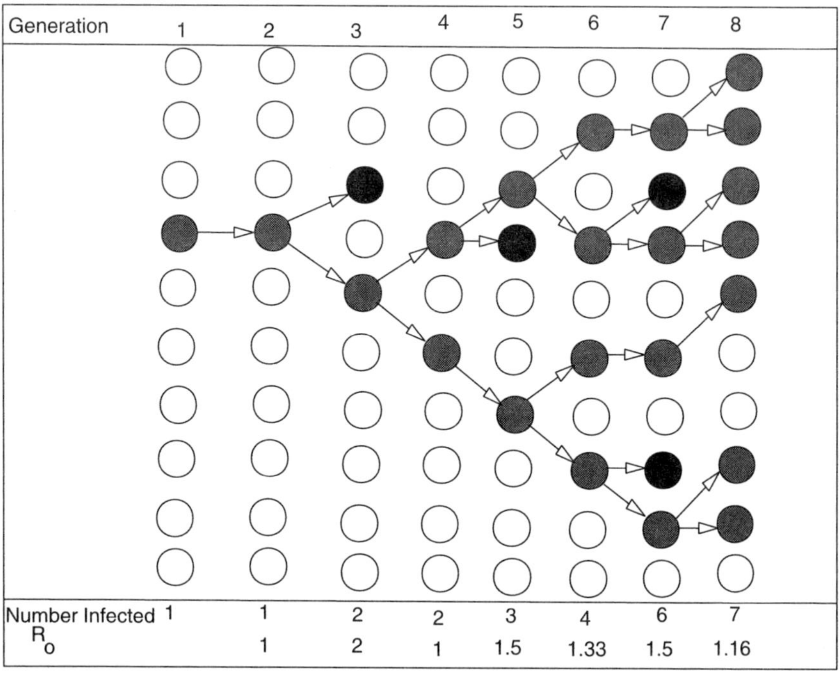

FIGURE 3.5 Diagrammatic illustration of chains of transmission between hosts.

An illustrative example is presented in Figure 3.6, which records a cross-sectional serological survey for antibodies to the measles virus, with an average age at infection of approximately 5 years.

Although these methods are easily applied to derive estimates of transmission intensity (R_0) and vaccine coverage required to block transmission, it is extraordinary how rarely they are used in public health practice. This may be due to a degree of mistrust of mathematical methods among public health scientists, perhaps related to the simplifications made in model construction. However, if the mathematical models are developed in close association with empirical measurement, and complexity is added slowly in a manner designed to promote understanding of which variables are key in determining the observed pattern, the methods can be very valuable in designing public health policy.

LIMITS OF MODELS

The real world is often much more complex than that portrayed in the simple assumptions that form a set of differential equations. Additional

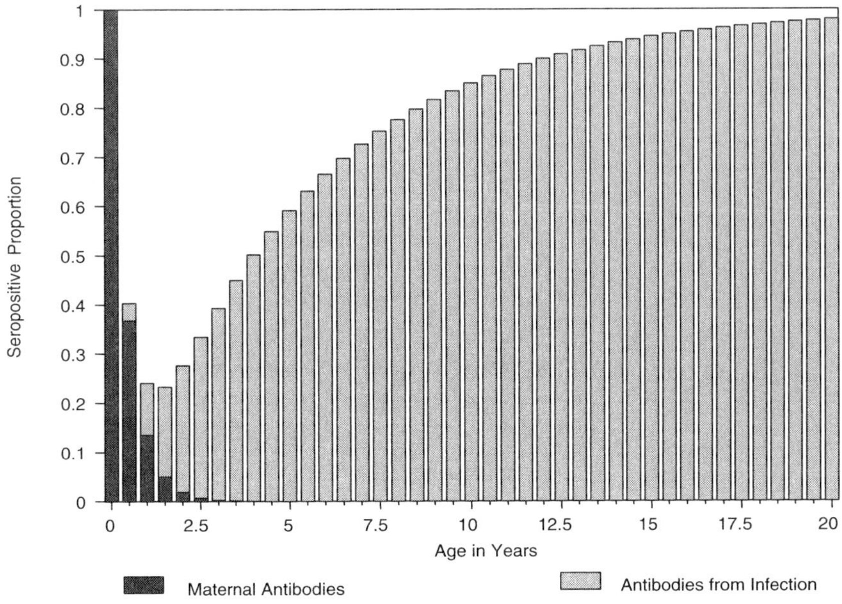

FIGURE 3.6 Age-stratified serological profile for measles virus infection.

sets of derivatives are typically needed to capture this complexity, beyond the rate of change with respect to time recorded, for example, in Figure 3.3. The rate of change with respect to age is important (see Figure 3.5), as is the rate of spread of an epidemic with respect to space. Including time, age, and space results in three derivatives within a set of partial differential equations, necessitating significant computational power to generate numerical solutions of the spatial diffusion, age distribution, and time course of an epidemic. Today, however, such approaches are possible, and the resultant models can be powerful predictive tools when used in conjunction with good biological and epidemiologic data that record the key parameters.

These methods can be used in a wide variety of circumstances involving much biological complexity. One illustration is their use to assess the potential value of a vaccine against the chicken pox virus (and the associated disease, zoster, in immunocompromised individuals and the elderly). The mathematical models used in analyzing this problem are both deterministic and stochastic and were able to show that cases of zoster in the elderly contributed approximately 7 percent of the total transmission intensity (i.e., primary infection plus transmission from zoster cases). Models were parameterized using data from small isolated island communities where it is possible to show that the virus would not be able to persist endemically without the help of transmission from zoster cases. This was informative from a public health perspective because it demonstrated that any immunization program targeted at the chicken pox virus and based on the immunization of young children would take a long time (i.e., many decades) to have its full impact because zoster in older people would still cause a degree of transmission among those not immunized.

Mathematical models have been applied most widely for the antigenically stable, vaccine-preventable childhood diseases. They have proved capable of predicting observed changes in both the average age of infection and the interepidemic period as a result of immunization programs. They have also proved reliable in predicting how a given level of immunization will change the herd immunity surface in a population. This surface represents changes in the proportion seropositive with respect to age and time (see Figure 3.7). Infection by the rubella virus provides a useful illustration of this predictive power. In this case the average age of infec-

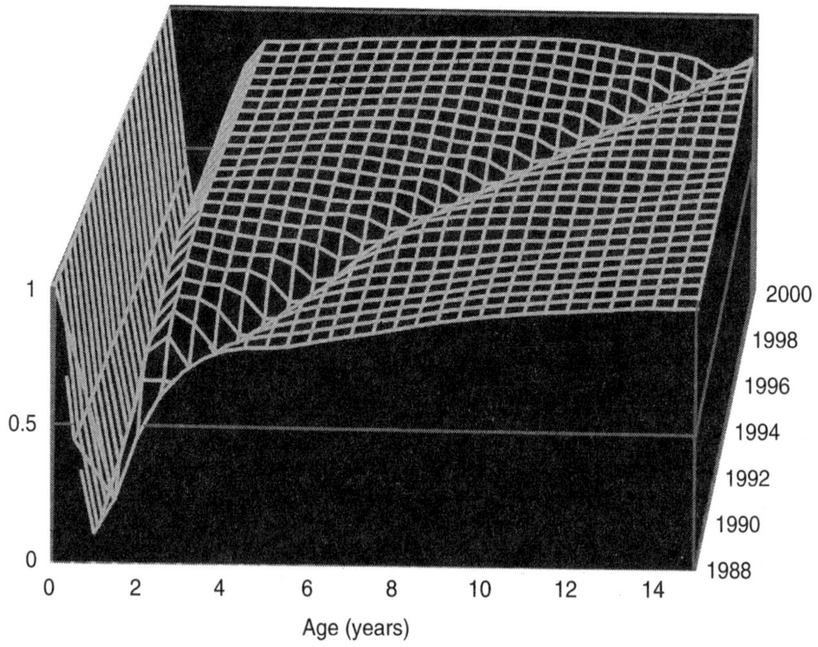

FIGURE 3.7 Herd immunity profile across age classes and through time.

tion was about 8 to 9 years prior to immunization in most developed countries. As we immunize a greater and greater fraction of the population, the age profile shifts, with the average age of infection increasing. It is important to understand that in a public health context this shift in the age distribution is not a failure of the immunization program but rather a natural consequence of the intervention. In the case of rubella this shift in the age distribution has important public health implications since infection during pregnancy can induce a serious disease—congenital rubella syndrome—in the unborn infant. Immunization programs targeted at the young must be at a sufficiently high level of coverage so as not to shift many young women into the pregnancy-age classes still susceptible to infection than was the case prior to the start of vaccination. Mass immunization induces a perturbation in the dynamics of the transmission system, which reduces the net force or rate of transmission of the infectious agent such that those who failed to be immunized have a lower exposure to the infectious agent. This is the so-called indirect effect of mass immunization. Normally, this is beneficial; however, in the case of rubella, infection at an older age carries an increased risk of serious morbidity (in this case for the unborn child of a pregnant women who contracts the infection).

CONSTRUCTING COST-BENEFIT MODELS

New techniques, such as saliva-based antibody tests, can facilitate the construction of herd immunity profiles for many infectious agents, including influenza. Using such epidemiologic data, model development can go one stage further and ask questions about vaccine efficacy and cost effectiveness of different control interventions. Virtually all published cost-benefit analyses in this area are incorrect because they typically only take into account the direct benefit of immunization in protecting the individual who is immunized. The indirect benefit is the herd immunity effect, as discussed in the preceding section, where those still susceptible experience a reduced rate of infection due to the immunity created by the vaccination of others. The percentage gain from the indirect effect of herd immunity is often quite dramatic, particularly at high vaccine uptake. This point is illustrated in Figure 3.8 for measles infection. It means that many immunization programs for the common vaccine-preventable infections are much more cost beneficial in terms of reducing morbidity and mortality than is currently appreciated.

Much of current methodology is designed for the antigenically stable infectious agents. Many important pathogens, particularly bacterial pathogens, exist as sets of antigenically related strains. Strain structure often varies between different communities and any significant change over time that is due to unknown selective pressures. Strain structure is best

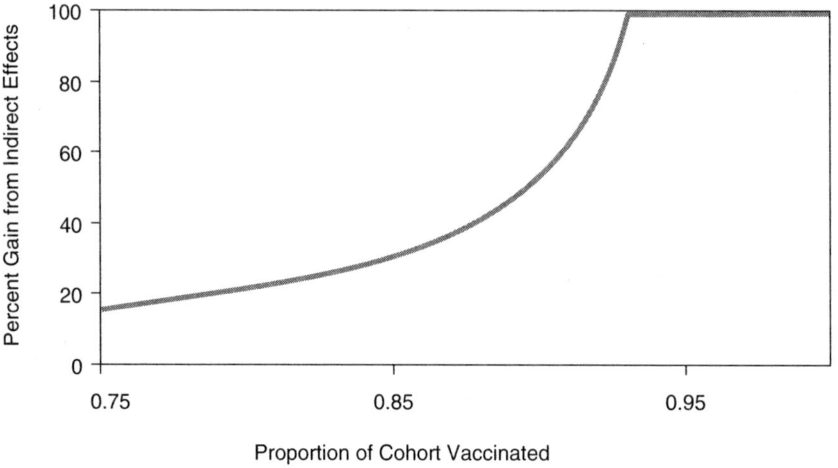

FIGURE 3.8 Percent gain in immunity from the indirect effects of herd immunity.

measured in terms of some easily identifiable phenotypic property. In the case of vaccine development, one such property would be recognition of the antigens of the pathogens recognized by the human immune system. Consider the case of a single dominant antigen coded for by a defined gene. If the pathogen expresses several epitopes on a particular gene and these epitopes are variable between strains, or where there is recombination taking place, the model must account for the dynamics of an evolving system in which different strains may have antigenically distinct combinations of epitopes. For example, if there were a whole series of phenotypes (= strains) that share some but not all alleles coding for the different epitopes, exposure to one strain may not confer immunity to another strain if that phenotype has a different set of alleles. If an organism shares all its alleles with another strain, the immune system is likely to confer complete protection against either strain from past exposure to either strain. If it shared only one out of, say, three alleles, there may be a degree of cross-protection but not complete protection.

These multistrain systems are dynamically very complex, and cycles plus chaos may be generated for fluctuations in prevalence over time by even simple models that represent cross-immunity generated by exposure to pathogens that share alleles. Much more research is required in this area to generate predictive tools to further our understanding of the likely impact of a new generation of vaccines targeted at antigenically complex bacterial pathogens, such as the pneumococcal organisms where there is a degree of cross-immunity between different strain types. It is important to understand these complex systems not only for the assessment of vaccine impact but also to understand fluctuations in strain structure over time in unimmunized populations.

THE CASE OF MAD COW DISEASE

Mad Cow disease (i.e., bovine spongiform encephalopathy [BSE]) is induced by a transmissible etiological agent in the form of an abnormal form of the prion protein. In the mid 1980s the disease developed as a major epidemic in cattle in the United Kingdom and it still persists in Great Britain but at a low and slowly decreasing rate of incidence (see Figure 3.9). The epidemic was created by the recycling of material from infected cattle in animal feeds given to cattle from the late 1970s through the 1980s. The concern for human health first arose in 1996 with the first report of a new strain of Creutzfeldt-Jakob disease (vCJD) in humans, which appeared to be related to BSE. Subsequent research has suggested that exposure to the etiological agent of BSE induces vCJD. The prion diseases can be diagnosed on the onset of characteristic symptoms, but no

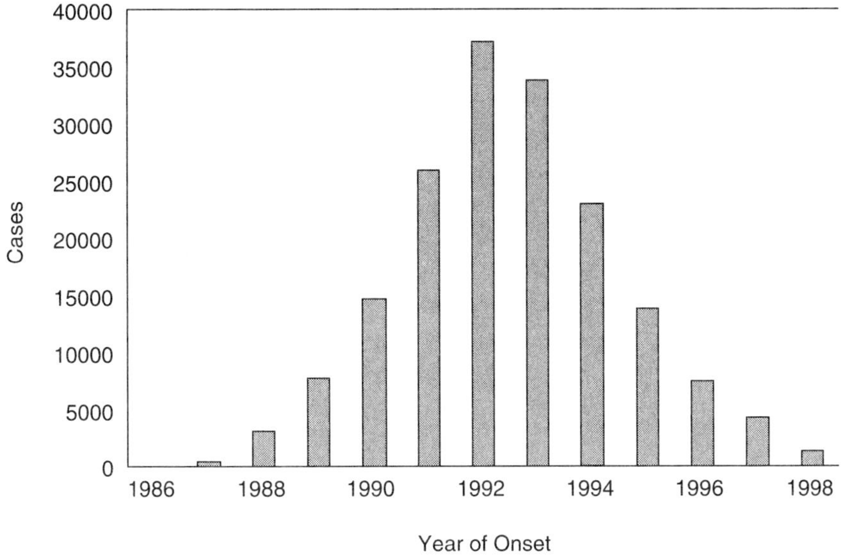

FIGURE 3.9 BSE epidemic in Great Britain (incidence/year).

reliable test exists at present to detect infection with the etiological agent prior to the onset of disease. The diseases are characterized by long incubation periods. In cattle the average incubation period is 5 years, while that of vCJD in humans is thought to be much longer.

Given a time course of the incidence of infection of BSE in cattle plus knowledge of the incubation period distribution, an important epidemiologic task is to estimate by back-calculation methods the numbers of infected animals that might have entered the human food chain via the consumption of infected beef or cattle products. These back-calculation methods were first used in the study of the AIDS epidemic to estimate the prevalence of HIV infection given observed time series of AIDS cases. The problem is more complicated in the case of BSE given the short life expectancy of cattle, which is on the order of 2 to 3 years from birth. This observation implies that most infected animals are slaughtered prior to the onset of symptoms of disease. Mathematical models that meld cattle demography with epidemiologic details of the incubation period distribution of BSE plus the major transmission route (horizontal and vertical) are required to assess the degree of exposure of the human population in Great Britain to the BSE agent.

Using such models, recent research has detailed the degree of exposure, which records temporal changes in the incidence of BSE, the incidence of new infections, and the prevalence of infected animals (Anderson et al., 1996). Taking the analysis further, to estimate the potential size and duration of the vCJD epidemic in humans is much more difficult due to the many unknowns surrounding this new disease. The major ones are the incubation period distribution and the infectivity of a unit quantity of contaminated beef. To complicate matters further, the density of abnormal prion protein in an infected cow changes significantly over the incubation period of the disease. It rises to very high levels in the brain and other neural tissues in the period immediately before the onset of symptoms of disease. Models can be constructed to make predictions of the possible future course of the epidemic, given the past history of exposure to BSE-contaminated material plus the observed time course of the incidence of vCJD. However, care must be taken to carry out many simulations, which embed methods for varying the values of the key unknown variables over ranges designed to cover sensible limits to these values. Such work requires considerable numerical analysis to generate prediction bounds on the future course of the epidemic. These bounds are very wide and do not eliminate either very small or very large epidemics. The major policy conclusion from this research is simply that the future is very uncertain and will remain so for many years. However, this conclusion is still useful since in its absence some scientists or policymakers may have been tempted to opt for one or other of the small or large epidemic scenarios and plan accordingly. The only sensible conclusion at present is that a large epidemic spread over many decades cannot be ruled out. The methods developed for the analyses of the BSE and vCJD epidemics may be of use in the future in studying the spread of other new infectious diseases.

EXPERIMENTAL SYSTEMS IN BIOLOGY

An exciting and developing area in the application of mathematics in biology and medicine is in understanding how the immune system interacts with the pathogen in an individual host. All the techniques that are used in epidemiologic population-based studies can be applied rigorously to the study of the population dynamics and evolution of pathogens in the human host. In addition, the opportunities for measurement and quantification of key parameters is in some sense much greater in this area due to new molecular and other methods for quantifying pathogen burden and associated immunological parameters such as cell life expectancies and rates of cell division.

One of the most important areas of application has been in HIV research, where a series of studies have used simple deterministic models to study HIV pathogenesis, viral life and infected cell life expectancies, the rate of evolution of drug resistance, and the dynamics of the immune system under invasion by the virus (e.g., Ho et al., 1995; Ferguson et al., 1999). Similar methods are now being used to study bacterial pathogens and other viral infections. The human immune system is very complex and nonlinear, especially in its response to replicating pathogens. There seems to be little doubt that mathematical models will play an important role in helping to unravel this complexity. Furthermore, melding this type of research with studies of the pharmacokinetics and pharmacodynamics of therapeutic agents designed to kill pathogens or inhibit their replication should provide new insights into the design of more effective therapies or the design of better ways to use existing drugs. A particular area of importance in this context will be the study of how best to slow the evolution of drug resistance, both for antivirals and antibacterial agents.

SUMMARY

Today, sophisticated mathematical models can be developed to estimate the size and speed of spread of an emerging epidemic, irrespective of the nature of the infectious agent or the size and spatial distribution of the affected population. These models not only can help in the development of policy options for containment but can also assist in the development of guidelines for effective population treatment programs.

REFERENCES

Anderson, R., and R. May. 1991. Infectious Diseases of Humans: Dynamics and Control. Oxford: University of Oxford Press.

Anderson, R. M., C. A. Donnelly, N. M. Ferguson, M. E. J. Woolhouse, C. Watt, et al. 1996. Transmission dynamics and epidemiology of BSE in British cattle. Nature, 382:779-788.

Ferguson, N. M., F. de Wolf, A. C. Ghani, C. Fraser, C. Donnelly, et al. 1999. Antigen driven $CD4^+$ T-cell and HIV-1 dynamics: Residual viral replication under HAART. Proceedings of the National Academy of Sciences, 21:96(26):15167-15172.

Ho, D., A. U. Neumann. A. S. Perelson, W. Chen, J. M. Leonard, and M. Markowitz. 1995. Rapid turnover of plasma virions and CD4 lymphocytes in HIV-1 infection. Nature, 373:123-126.

Nissinen A, M. Leinonen, P. Huovinen, E. Herva, M.L. Katila, S. Kontiainen, O. Liimatainen, S. Oinonen, A.K. Takala, P.H. Makela. 1995. Antimicrobial resistance of *Streptococcus pneumoniae* in Finland, 1987-1990. Clinical Infectious Diseases, 20(5):1275-1280.

Schwartländer, B., G. Garnett, N. Walker, and R. Anderson. 2000. AIDS in the new millennium. Science, 289:64-67.

4

Expanding the Worldwide Influenza Surveillance System and Improving the Selection of Strains for Vaccines

Nancy J. Cox

INTRODUCTION

The Centers for Disease Control and Prevention (CDC) and other groups participating in the World Health Organization's (WHO) global influenza surveillance network track the ever-changing influenza viruses that infect humans. But influenza viruses have characteristics that make this task difficult. One of these characteristics is their ability to evolve relatively rapidly. Another factor is their ability to infect avian and other mammalian species and occasionally cross host species' barriers and evolve rapidly in a new host. Expanding global influenza surveillance, in conjunction with automating key laboratories and establishing an informatics infrastructure, could dramatically improve our ability to track influenza viruses and prevent and control influenza outbreaks in the future.

INFLUENZA MORTALITY AND MORBIDITY DURING PANDEMIC AND INTERPANDEMIC PERIODS

The most widely recognized consequence of influenza is the morbidity and mortality due to pandemics (worldwide epidemics). For example, more than 500,000 excess influenza-related deaths occurred in 1918 in the United States during the "Spanish" influenza pandemic alone, and a total of more than 600,000 deaths have been attributed to the Spanish, Asian (1957), and Hong Kong (1968) influenza pandemics combined. Furthermore, the morbidity due to pandemics is well recognized to cause considerable social disruption and substantial economic losses. It is less well recognized that the cumulative totals for mortality and morbidity are actually much greater during interpandemic periods than pandemic years. For just the quarter century between 1970 and 1995, there were also more than 600,000 excess influenza-related deaths. During the current interpandemic period, an annual average of 20,000 excess influenza-related deaths and more than 100,000 excess influenza-related hospitalizations have occurred since 1973. In addition, serious epidemics have resulted in more than 40,000 influenza-related deaths and 200,000 influenza-related hospitalizations.

IMPACT OF INFLUENZA ON AN AGING POPULATION

Influenza affects different age groups in different ways. During the current interpandemic period, more than 90 percent of the excess mortality from pneumonia and influenza occurs in the elderly. Thus, the aging of the population, one of the demographic changes occurring worldwide, has significant public health implications with regard to influenza. Between 1985 and 2025, the population age 65 years and older in the United States will double. In China and India this population group will triple in size. If effective influenza prevention and control measures are not instituted, the world will suffer from substantial societal and economic costs related to influenza, particularly among older individuals. Although mortality during the current interpandemic period occurs primarily among those 65 and older, influenza-related hospitalizations occur in both the very young and the elderly. School-aged and younger children have the highest rates of medically attended influenza-related illness.

During the three pandemics of influenza that occurred during the twentieth century, the pattern of mortality differed from that seen today. During the Spanish influenza pandemic of 1918, more than 99 percent of excess pneumonia and influenza deaths occurred among persons under age 65. However, as the new pandemic viruses became established in the human population and continued to circulate and evolve, a decrease was observed in the proportion of influenza-related deaths occurring in those under age 65. A similar pattern was observed for the pandemics of Asian influenza in 1957 and the Hong Kong influenza in 1968. When the Asian strain of influenza first emerged in humans, approximately 40 percent of influenza-related deaths occurred in those under age 65. During the 1968 pandemic, approximately 60 percent of the deaths occurred in this age group. However, smaller proportions of influenza-related deaths were observed among those under 65 years of age, as these two novel viruses continued to circulate among humans and population immunity developed.

WHO'S GLOBAL INFLUENZA SURVEILLANCE NETWORK

Influenza exhibits a complex seasonal pattern of circulation. In temperate regions of the Northern Hemisphere, influenza activity usually peaks during the months of December, January, and February. In temperate regions of the Southern Hemisphere, the peak of influenza activity occurs from June through August. In tropical regions, influenza can be isolated almost year round, and in some areas there are two peaks of influenza activity. This pattern of influenza circulation keeps WHO's global influenza surveillance network busy throughout the year. This network, established by the WHO over 50 years ago, has grown to include four WHO collaborating centers for influenza (CCIs) located in London, Atlanta, Melbourne, and Tokyo. It also includes approximately 110 national influenza centers (NICs) in over 80 countries worldwide. These centers are fairly well distributed around the world; however, some countries in Africa, South America, and the Middle East currently do not participate in WHO's global influenza surveillance network.

Active NICs isolate influenza viruses and identify them using a WHO influenza reagent kit that is prepared annually at the CDC and distributed throughout the world. Subsequently, the isolated viruses are sent to one or more CCIs to be analyzed in greater detail. The purpose of this virological surveillance system is to detect a variant strain that could cause the next epidemic or pandemic of influenza with sufficient early warning to include the new strain in influenza vaccine. Epidemiologic information is also collected and sent to WHO in an effort to document the extent of influenza activity during the time the viruses were isolated. In addition, the United States (and a number of other countries) has its own domestic network, which includes more than 70 WHO collaborating laboratories, many of which are located in state health departments.

The four CCIs have several key responsibilities, including the following: analyzing influenza viruses from around the world; communicating with the NICs; providing training; communicating with other WHO collaborating centers; preparing a WHO influenza reagent kit; and providing data for WHO's annual vaccine recommendations. All four collaborating centers use antigenic and genetic techniques to comparatively analyze the influenza viruses received. The CDC has also provided extensive support for laboratory training. For example, in 1994 and 1999 training in Beijing was provided for participants from municipal and provincial antiepidemic stations across China. The CCIs communicate with each other, with the NICs, and with WHO headquarters in Geneva. Much of this information is published in WHO's *Weekly Epidemiological Record.*

CHOOSING INFLUENZA VACCINE STRAINS

The four WHO collaborating centers for influenza are responsible for providing data for the influenza vaccine recommendations that WHO issues to the world twice a year—in February for the Northern Hemisphere and in September for the Southern Hemisphere. These recommendations are based on antigenic, genetic, serological, and epidemiologic data.

After viruses are received by the CCIs, they are amplified by growth in tissue culture or embryonated hens' eggs. A hemagglutination inhibition (HI) test—a serological assay that can rapidly screen a large number of viruses—is used to detect new antigenic variants. This test is performed using postinfection ferret sera made against reference strains. These sera are very sensitive for detecting antigenic differences among viruses. Low reactors, that is, those viruses that are not well inhibited by the reference ferret antisera and appear to be variant viruses, are retested, and viruses that are confirmed to be variants are inoculated into ferrets to produce a corresponding antiserum for further HI testing.

The WHO relies on ministries of health and the NCIs to report high levels of influenza-like activity coincident with isolation of the variant strains. Before selecting strains for inclusion in the vaccine, the WHO also determines if a reduced postvaccine immune response to variant viruses occurs in individuals who have received the current vaccine. After all of this information is synthesized, recommendations are made regarding the specific viruses that should be included in the next year's influenza vaccine. Once the vaccine strain recommendations are issued, suitable candidate vaccine strains are distributed to the vaccine manufacturers by the WHO collaborating centers or by regulatory authorities.

In 1985 the CDC began prospective sequencing of hemagglutinin (HA) genes to determine if the data generated would be useful for vaccine strain selection. In recent years the CDC has sequenced the HA genes of many influenza field strains, particularly those that may have an altered pattern of reactivity in the HI test. CDC has focused on the HA gene because antibodies to hemagglutinin determine whether or not a person is protected against infection by influenza. When CDC began prospective sequencing of the HA genes of influenza field strains it was not known if sequencing would be a useful adjunct to reference serological analysis for vaccine strain selection; however, it is now well established that molecular methods such as sequencing are extremely useful. Sequence analysis of HA genes can provide information concerning the molecular basis for antigenic drift, an area where more information is needed. We calculate rates of change for the HA at both the nucleotide and amino acid levels and also examine the types of amino acid changes in the hemagglutinin that appear to confer antigenic changes and where those changes are located in the three-dimensional structure of the molecule. We are also interested in learning if there are predictable patterns of change. Information gained from ongoing sequencing indicates that the number of pathways for evolution of the virus may be limited and that there may be some predictability in the patterns of change. It may be possible to identify specific clues in the sequence data and devise methods to help predict which strains are most likely to cause epidemics in the future. The CDC has also been working with scientists at the Los Alamos National Laboratory to establish an international database for influenza gene sequence data.

Table 4.1 shows the "WHO report card," indicating the degree of antigenic relatedness or "match" between the epidemic strains that have circulated in the United States over the past 10 years and the correspond-

TABLE 4.1 Match Between Epidemic and Vaccine Strains of Influenza, USA, 1989 to 2000

Season	Epidemic Strain(s)	Vaccine(s)	Antigenic Match
1989 to 1990	A/Shanghai/11/87(H3N2)	A/Shanghai/11/87(H3N2)	++++
1990 to 1991	B/Hong Kong/22/89	B/Yamagata/16/88	+++
1991 to 1992	A/Beijing/353/89(H3N2)	A/Beijing/353/89(H3N2)	++++
1992 to 1993	B/Panama/45/90	B/Panama/45/90	++++
	A/Beijing/32/92(H3N2)	A/Beijing/353/89(H3N2)	+
1993 to 1994	A/Beijing/32/92(H3N2)	A/Beijing/32/92(H3N2)	++++
1994 to 1995	B/Panama/45/90	B/Panama/45/90	+++
	A/Shangdong/09/93	A/Shangdong/09/93	++++
1995 to 1996	A/Texas/36/91(H1N1)	A/Texas/36/91(H1N1)	++++
	A/Johannesburg/33/94(H3N2)	A/Johannesburg/33/94(H3N2)	++++
1996 to 1997	A/Wuhan/359/95(H3N2)	A/Wuhan/359/95(H3N2)	+++
1997 to 1998	A/Sydney/05/97(H3N2)	A/Wuhan/359/95(H3N2)	+
1998 to 1999	A/Sydney/05/97(H3N2)	A/Sydney/05/97(H3N2)	++++
1999 to 2000	A/Sydney/05/97(H3N2)	A/Sydney/05/97(H3N2)	++++

NOTE: +, some cross-reaction; ++, moderate cross-reaction; +++, substantial cross-reaction; ++++, identical or minimal difference.

ing vaccine strains. This table shows that despite a good record overall there was a poor match between the vaccine strain and the circulating strains during the 1997 to 1998 influenza season. Because the A/Sydney/5/97 variant was not detected until the summer of 1997, there was not enough time to include it in the vaccine before the following influenza season. Despite this mismatch, the similarity between the antigenic properties of the vaccine strains and those of the predominant circulating viruses has improved during the past 10 years compared to previous decades. Factors that have contributed significantly to an improved match include better worldwide surveillance, especially in some European countries; the expansion of influenza surveillance in China; and the use of molecular analysis as an adjunct to serological analysis.

Over the past 10 or 15 years, influenza vaccination rates have improved dramatically in certain segments of the U.S. population. Vaccination rates among people 65 and older have increased from approximately 33 percent in 1989 to 63 percent in 1997. Thus, the Healthy People 2000 Public Health Service objective of vaccinating 60 percent of this population was reached before the year 2000. However, vaccination rates have remained low among people under age 65 with high-risk conditions; more effort must be focused on improving vaccine coverage in this group.

LIMITING FACTORS FOR SPECIAL INVESTIGATIONS AND FOR ANNUAL VACCINE STRAIN SELECTION

WHO collaborating centers for Influenza face a number of limiting factors that affect vaccine strain selection as well as special investigations like those that occurred in Hong Kong after human infections with avian influenza A (H5N1) and (H9N2) viruses were documented. During the H5N1 investigation, it was necessary to test more than 3,000 sera using multiple tests and to repeat many assays. Development and modification of confirmatory assays were essential components of this investigation. Also, the entire genomes of the H5N1 influenza viruses isolated from humans needed to be sequenced quickly in order to try to determine why these viruses were able to jump the host-species barrier. Automation of these processes would have aided this work.

There are also limiting factors for generating data for influenza vaccine strain selection. Influenza surveillance is not uniform throughout the world, and in some countries there is a need to establish influenza surveillance or to increase the number of influenza viruses that are isolated. For example, the WHO network in certain parts of the world, including Africa, needs to be expanded. The CDC has been focusing on improving influenza surveillance in China, Vietnam, and Russia; influenza surveillance is also being enhanced in several countries in Asia, South America, and Europe. In countries with very active NICs, many influenza isolates are often identified; however, not all of them are analyzed in detail. There is a need to increase the number of isolates sent to WHO CCIs for detailed analysis. This means that CCIs need to increase their capacity to grow, store, and analyze influenza strains and to devise improved methods to handle high throughput of influenza viruses at critical times. For example, in February, when the vaccine strain selection for the Northern Hemisphere is being made, and in September, when the vaccine strain selection for the Southern Hemisphere is being made, it would be extremely useful to have better ways to analyze large numbers of influenza isolates rapidly. There is also a need to expand the number of isolates sequenced and the number of genes for each virus. Furthermore, it would be extremely useful to extend serological testing of isolates by using a functional assay such as a neutralization test.

CONCLUSION

Routine surveillance efforts at several CCIs have reached maximum capacity, unless staffing is expanded significantly or processes are automated. Automation would greatly increase the number of influenza isolates and serum specimens that could be analyzed and would also increase the numbers and types of tests that could be performed. The new data generated would be extremely useful for extending our knowledge of the epidemiology, ecology, evolution, and pathogenesis of influenza viruses as well as for influenza vaccine strain selection. Efforts should be made to introduce automation into one or more of the WHO CCIs so that these additional data can be generated rapidly.

5

Addressing Emerging Infectious Diseases, Food Safety, and Bioterrorism: Common Themes

James M. Hughes

Infectious diseases are a leading cause of death worldwide and the third leading cause of death in the United States. According to the World Health Organization (1998), of the 52 million deaths that occurred globally in 1997, one-third, or 17 million, were attributable to infectious disease. The leading killers are acute lower respiratory tract infection (primarily pneumonia), tuberculosis, diarrhea and dysentery, HIV infection and AIDS, malaria, measles, and hepatitis B. Many of these diseases have a propensity for drug resistance, which contributes to some of the mortality. In the United States, infectious disease mortality has been on the rise, largely due to the rapid spread of HIV but also due to pneumonia and bloodstream infection. Despite these data, because of improvements in sanitation and the availability of effective antibiotics and vaccines, considerable complacency has developed over the past 30 years in the United States regarding infectious diseases.

More recently, infectious disease outbreaks have presented major challenges and have reminded us that we live in a global village. These outbreaks have included:

- Plague in India.
- Ebola hemorrhagic fever in Zaire and Gabon.
- Leptospirosis in Nicaragua and the United States.
- *Cyclospora* gastroenteritis in the United States and Canada.
- *Escherichia coli* O157:H7 hemorrhagic colitis in Japan and the United States.

- *Staphylococcus aureus* infections caused by strains with diminished susceptibility to vancomycin in Japan and the United States.
- H5:N1 influenza in Hong Kong.
- Rift Valley fever in Kenya, Somalia, and Saudi Arabia.
- Nipah virus infection in Malaysia and Singapore.
- Foodborne salmonellosis, shigellosis, and listeriosis in the United States.
- West Nile encephalitis in New York City and nearby areas.

Addressing the major challenges posed by these outbreaks requires timely epidemiologic investigations, use of sophisticated laboratory techniques, and effective communication of results and recommendations for prevention. We cannot afford to ignore infectious disease problems in other parts of the world. From our experience at the Centers for Disease Control and Prevention (CDC) in dealing with a number of these emerging and reemerging infection outbreaks over the past 6 to 7 years, several important and recurrent lessons have emerged. These outbreaks emphasize the importance of surveillance, the need to have the ability to conduct prompt epidemiologic investigations, and the need to have available critically important public health laboratory capacity. In addition, human resource development through training is necessary to strengthen prevention and control programs at all levels.

A 1992 LANDMARK REPORT

In 1992 a landmark Institute of Medicine (IOM) report, *Emerging Infections: Microbial Threats to Health in the United States*, highlighted this complacency, emphasized the threats posed by infectious diseases, identified the factors that contribute to disease emergence and reemergence, and stressed the need to heighten vigilance and strengthen disease detection and response capacity. Notably, the report provided the following definition of emerging infections: "New, re-emerging, or drug-resistant infections are those whose incidence in humans has increased within the past few decades, or whose incidence threatens to increase in the near future." Notice that this definition is very broad and that it specifically includes drug resistance. The report identified six major factors that contribute to disease emergence and reemergence:

- Changes in human demographics and behavior.
- Advances in technology and industry.
- Economic development and changes in land-use patterns.
- Dramatic increases in the volume and speed of international travel and commerce, including foodstuffs and animals.

- Microbial adaptation and change in response to selective pressures in the environment.
- Breakdown in public health measures.

The IOM report could not have been timelier. Within 6 months of its publication, clinicians, microbiologists, and public health officials in the United States were confronted with three dramatic examples of infectious disease threats. The first was an interstate foodborne disease outbreak of hemorrhagic colitis and hemolytic uremic syndrome caused by *E. coli* O157:H7 linked to undercooked hamburger served by a fast food restaurant chain. The second event was the largest waterborne disease outbreak in U.S. history, which resulted in more than 400,000 cases of cryptosporidiosis in Milwaukee, Wisconsin. The third event was an outbreak of acute respiratory distress syndrome caused by a previously unrecognized hantavirus in the southwestern United States. Simultaneously, the incidence of nosocomial infections caused by vancomycin-resistant enterococci and that of penicillin resistance in community-acquired pneumococcal infections caused by penicillin-resistant strains have increased.

THE CDC RESPONSE

Because more than half of the recommendations in the 1992 IOM report were directed at the CDC, that agency developed a strategy for addressing emerging infections in consultation with outside experts in the areas of clinical infectious diseases, microbiology, and public health. The strategy contains four goals that focus on strengthening surveillance and response capacity; addressing applied research priorities; strengthening the public health infrastructure at the local, state, and federal levels; and improving prevention and control programs. Incremental implementation of this strategy began in 1994, and an updated version was issued in September 1998 (CDC, 1998).

Two initiatives under way to incrementally implement this plan are worth mentioning. First, the CDC has organized and provided support for nine emerging-infections programs. These are financial awards made competitively to state health departments to develop consortia within their states, reaching beyond the traditional public health system to involve the clinical and academic communities in these projects.

In addition, the CDC has provided support to 43 additional jurisdictions, including 39 state public health departments and four large cities—Los Angeles, New York City, Houston, and Philadelphia—to help them repair some of their deteriorated public health laboratory capacity and to provide resources for strengthening surveillance capacity, particularly via molecular fingerprinting. Without this molecular tool, the ability of public health departments to recognize, intervene, and control many outbreaks is severely limited.

Because of the importance of molecular fingerprinting, the CDC has begun to develop an integrated foodborne disease surveillance system in the United States. This national network, called PulseNet, involves four regional laboratories as well as the U.S. Department of Agriculture, the Food and Drug Administration, CDC laboratories, and many state public health laboratories that perform molecular fingerprinting. This effort has been critical in recognizing a number of regional and national foodborne disease outbreaks in the past three years.

THE NEED TO ADDRESS ANTIMICROBIAL RESISTANCE

Antimicrobial resistance is a critically important clinical and public health issue, especially since the drug development pipeline has largely dried up. Dealing with the problem of antimicrobial resistance is complicated and requires a multifaceted strategy that starts with surveillance and requires rational use of antimicrobial agents, which relies on the availability of rapid diagnostic tests. Physicians need office-based tests to help them determine whether a patient has a bacterial or a viral infection and, if bacterial, whether it is drug susceptible or drug resistant. Regarding the crux of the problem of antimicrobial resistance, however, John Burke wrote a few years ago: "Despite the multifactorial nature of antibiotics resistance, the central issue remains quite simple. The more you use it, the faster you lose it" (Burke, 1995).

For many years professional societies have developed a number of recommendations for clinicians in an effort to improve antimicrobial usage. A recent publication developed by the CDC and the American Academy of Pediatrics provided family practitioners and pediatricians with some principles for judicious use of antimicrobial agents for pediatric upper respiratory tract infections. However, the public needs to be brought into this campaign against antimicrobial resistance to begin to understand that antimicrobial agents are important resources that must be conserved and used appropriately.

RESPONDING TO BIOTERRORISM

The bombings of the World Trade Center in New York and the federal building in Oklahoma City have demonstrated that we are vulnerable to terrorist attacks, adding a new dimension to the challenges posed by infectious diseases. Biological attack, which in the past was considered unlikely, now seems entirely possible, given that information on how to prepare such weapons is widely available and that actions have been carried out by groups such as Aum Shinrykyo, which released nerve gas in Tokyo's subway system and experimented with biological weapons.

Four components of the public health response to disease outbreaks are important regarding preparedness in addressing acts of bioterrorism. First, since initial disease detection is likely to take place at the local level, it is essential to work with the medical community, including emergency departments, infection control practitioners, poison control centers, and emergency responders. A 1998 report of the National Research Council on chemical and biological terrorism, recommends expanding the CDC's emerging-infections initiative to improve state and local surveillance infrastructure for detecting naturally occurring outbreaks as well as those potentially resulting from bioterrorism.

The second component is investigation and response, activities that are also likely to take place initially at the local level. Third, rapid diagnosis will be critical so that prevention and treatment measures can be implemented quickly. Because the agents that are most likely to be used as bioweapons are not currently major public health problems in the United States, biocontainment laboratory space and surge capacity to investigate and respond to outbreaks have been limited. In addition, future bioweapons events could involve organisms that have been genetically engineered to increase their virulence, manifest antibiotic resistance, or evade natural or vaccine-induced immunity. Finally, communication is crucial, as delays will increase the probability that more people will be exposed. Preparedness will require effective partnerships with local, state, and other federal public health agencies; clinicians and clinical microbiologists; research institutions; and industry. Preparedness for bioterrorism requires that these partnerships be extended to include the emergency response and law enforcement communities.

SUMMARY

Infectious diseases are important, evolving, and complex public health problems. Their prevention and control require application of sophisticated epidemiologic, laboratory, statistical, behavioral, and informatics approaches and technologies. An integrated approach involving epidemiologic, laboratory, behavioral, and information sciences is critical to the prevention and control of infectious diseases. A strong and flexible public health infrastructure is the best defense against any disease outbreak, whether naturally occurring or intentionally caused.

REFERENCES

Burke, J. P. 1995. How to maintain the miracle of antibiotics. Lancet, 345:977.

Centers for Disease Control and Prevention (CDC). 1998. Preventing Emerging Infectious Diseases: A Strategy for the 21st Century. Atlanta: CDC (http://www.cdc.gov/ncidod/emergplan).

Institute of Medicine. J. Lederberg, R. E. Shope, and S. C. Oaks, eds. 1992. Emerging Infections: Microbial Threats to Health in the United States, Washington, D.C.: National Academy Press.

National Research Council and Institute of Medicine. 1998. Improving Civilian Medical Response to Chemical or Biological Terrorist Incidents. Washington, D.C.: National Academy Press.

World Health Organization (WHO). 1998. The World Health Report 1998: Life in the 21st Century: A Vision for All. Report of the Director General. Geneva: WHO.

6

Laboratory Firepower for AIDS Research

Scott P. Layne and Tony J. Beugelsdijk

INTRODUCTION

HIV poses enormous challenges from a variety of directions. In less than three decades, it has grown from an unknown pathogen to a pandemic disease, chronically infecting more than 50 million people and killing nearly 3 million people each year. By the beginning of twenty-first century, an estimated 1 percent of the world's population will be afflicted, and end-stage AIDS will rank as one of the top five causes of death by infectious disease (Murray and Lopez, 1997). In response to this catastrophe, pharmaceutical companies are manufacturing new antiviral drugs (e.g., reverse transcriptase and protease inhibitors) that extend peoples' lives and investigators are examining chemokine receptors (e.g., CCR and CXCR) and cytotoxic cellular epitopes that offer promising leads for therapeutics and vaccine development (Baggiolini, 1998; Heilman and Baltimore, 1998). Despite such rapid progress, however, the fact remains that individual investigators tend to shy away from important problems dealing with enormous "experimental spaces" simply because they lack the necessary tools to move ahead.

The current models of discovery and follow-up are based on investigator-initiated research and relatively small-scale efforts, but in waging the war on AIDS the research community should also consider how intermediate-scale research infrastructures (first referred to as "collaboratories" by William A. Wulf) could feasibly contribute (National Research Council, 1993). Basic as well as clinical investigators consequently should ask three questions about the potential roles for such collaboratories. First, what activities in AIDS research (see Box 6.1) require enormous inventories of data and information? Second, what categories of laboratory-based tests (see Figure 6.1) must be carried out to create such inventories? Third, what technologies are available for creating high-throughput research facilities that are cost effective, reliable, and flexible? To move forward with collaboratory-based efforts for AIDS research, there must be some form of consensus among investigators and administrators. After all, scientific and medical research requirements must drive the demand for high-throughput resources, not vice versa.

If sufficient agreement does indeed exist, a critical number of available technologies and scientific disciplines could certainly be brought together to level the playing field against the growing threat of HIV infection and AIDS mortality. The available tools would include reliable laboratory tests (methods), automation and robotic equipment (hardware), object-oriented programming languages (software), relational databases (informatics), shipping services (virtual warehousing), and Internet providers (communications). This paper sets forth, with such tools in hand, a blueprint for creating flexible laboratory and informatics facilities that could be operational within 2 to 3 years for basic science and a variety of clinical trials and public health efforts. As described later, important roles for such laboratories could include development of HIV vaccines and

BOX 6.1
Important Problems in Basic and Clinical AIDS Research that Could Benefit from Newer Models of High-Throughput Laboratory Research.

- Vaccine trials involving many volunteers, multiple arms, polyvalent immunogens, or strain-specific immunity.
- Therapeutic trials involving multiple antiretroviral drug regimens.
- Individualizing antiretroviral drug therapies with results from viral load measures, molecular sequencing assays, and drug susceptibility tests.
- Basic science investigations involving combinatoric assays with antibodies, cytokines, and chemokines.
- Epidemiologic and natural history investigations involving the isolation and characterization of many viral quasi species.
- Molecular biology investigations involving systematic mutational analysis of viral proteins.
- Surveys seeking correlations between HIV genotypes, phenotypes, and epidemiologic characteristics.

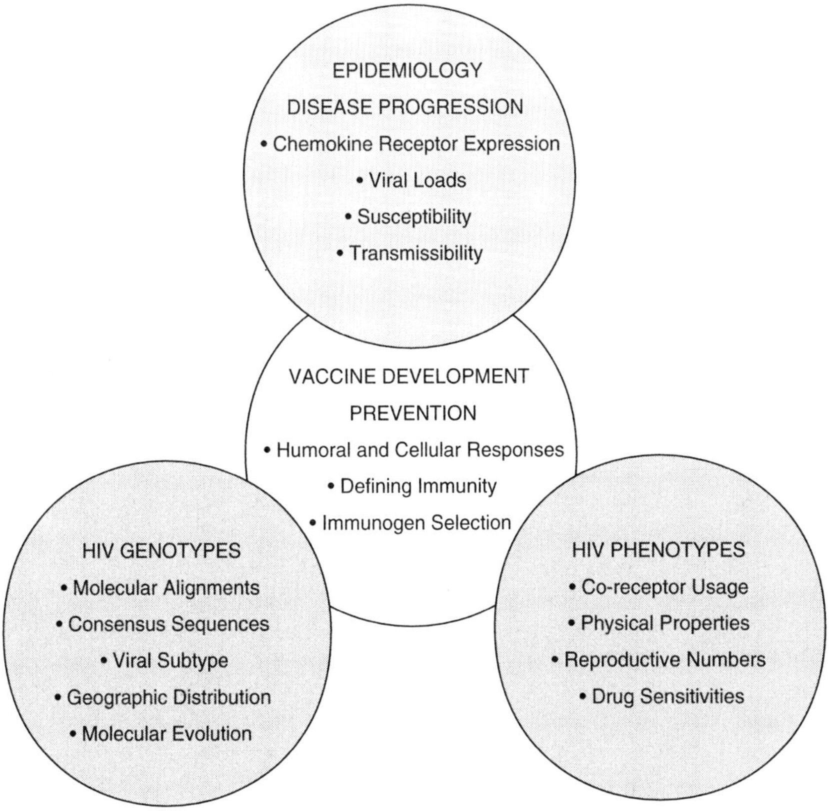

FIGURE 6.1 Throughout the world, HIV-infected individuals harbor an extremely broad array of wild-type viruses, and AIDS investigators have limited systematic information on these pathogens. The lack of basic knowledge of how viral reproduction, genetic mutation, and immune selection are interrelated raises the issue of whether comprehensive surveys should be initiated. Such surveys would build inventories of epidemiologic and laboratory-based data and seek correlations (if any) between phenotypic expressions, genotypic variations, and disease progression. The figure shows typical measures that apply to each category of data. Ultimately, it may be possible to identify the "right" assortments of HIV genotypes and phenotypes for vaccine development. Logical considerations would include the most prevalent and representative strains, but there is no clear consensus on this issue at present. An inventory of systematic information could help.

optimization of therapeutic regimens by investigators on national and international scales.

CURRENT LIMITATIONS

Unprecedented digital technologies are now at the fingertips of scientific and medical investigators in many geographic locations. To illustrate this point, consider the magnitude of three representative benchmarks from supercomputing, information storage, and worldwide communications. At many advanced computer facilities, numerical performances are surpassing $\sim 10^{12}$ floating point operations per second (teraflop speeds), database storage facilities are exceeding $\sim 10^{15}$ bits per site (petabit capacities), and Internet communication rates are achieving $\sim 10^{9}$ bits per second (gigahertz bandwidths). Despite such fantastic digital firepower, many important AIDS research efforts currently are limited by sheer physical firepower. The biggest rate-limiting step is the ability to perform vast numbers of tests in the laboratory (Layne and Beugelsdijk, 1998).

Most laboratory experiments in AIDS research are performed by human hands, usually in combination with sprinklings of labor-saving devices and small islands of automation. This semimanual approach may certainly lead to insights and breakthroughs, but such laboratory work is often repetitive and exceedingly tedious. Even with an army of postdoctoral students and laboratory technicians, human hands still remain the limiting factor in generating the vast quantities of raw information required for solving complicated problems. Fortunately, rather impressive technological advancements have been made in recent years, and so the time is now right for overcoming such manpower-related obstacles.

Virologists, immunologists, and molecular biologists have developed a variety of laboratory-based assays that are reproducible and readily adaptable to large-scale efforts. Engineers have developed innovative automation and robotics technologies that are capable of skyrocketing the number and variety of laboratory experiments. Computer scientists have developed Internet programming languages and database management systems that are providing the basic building blocks for improved environments in scientific collaborations. And physicists (driven by the need to share gargantuan amounts of data generated in high-energy particle physics experiments at a few large accelerator facilities) have developed the World Wide Web. Today, these developments are literally transforming the ways in which scientific collaboration and information distribution are taking place (Lanier et al., 1997). As described below, the integration and refinement of these capabilities hold significant promise for accelerating AIDS research in basic science, clinical trials, and public health efforts throughout the world. The objective is to create a high-throughput laboratory that is practical, flexible, and easy to use from any geographic location.

BATCH SCIENCE

The Internet can be used by laboratory-based investigators in either of two ways—for handling a series of *real-time* or *nonreal-time* operations. Real-time operations require specialized communications protocols that are redundant and failsafe. An actual illustration of real-time manipulations is access to a multiuser scanning electron microscope from a distance. The user operates the microscope's controls in one location, and the sample is manipulated and viewed while residing in another place (Chumbley et al., 1995). In general, few specialized applications in science and medicine require real-time operations from afar. Moreover, the Internet's hard wiring and protocols are not ready for handling large volumes of real-time manipulations every day, which will require much larger bandwidths (Germain, 1996).

On the other hand, nonreal-time operations do not require specialized communications protocols that are redundant and failsafe. Most scientific and medical research efforts can be supported by nonreal-time operations that are scripted with flexible software tools—an approach that will be referred to as *batch science via the Internet* (see Figure 6.2). Batch science machines would perform the manipulative work of hundreds of humans, serve as programmable laboratory technicians, and help AIDS investigators tackle certain types of big problems (Box 6.1).

A feasible illustration of batch science would be the undertaking of major AIDS vaccine trials (perhaps in the near future) with the assistance of an automated reference laboratory. In this example, investigators would collect numerous clinical samples from vaccinees before and after receiving the series of immunizations (i.e., the *efficacy* cohort) and, in a parallel effort, from HIV-infected individuals in close geographic proximity but not receiving such shots (i.e., the *carrier* cohort taking other forms of therapies). Investigators would then use flexible software tools to design and script the necessary laboratory tests while residing at their home bases. The available "toolbox" of laboratory tests would include full assortments of viral infectivity, molecular sequencing, and cytotoxic T lymphocyte assays (see Box 6.2). Next, investigators would airfreight the frozen samples to their collaborating facility (in another geographic location) and within just a few days the desired assays would be set up and carried out by high-throughput automation in accordance with the digital "assay scripts" that arrived via the Internet. Upon completing the necessary tests, data from the various assays would be electronically deposited into the trial's supporting database, and collaborating teams of investigators

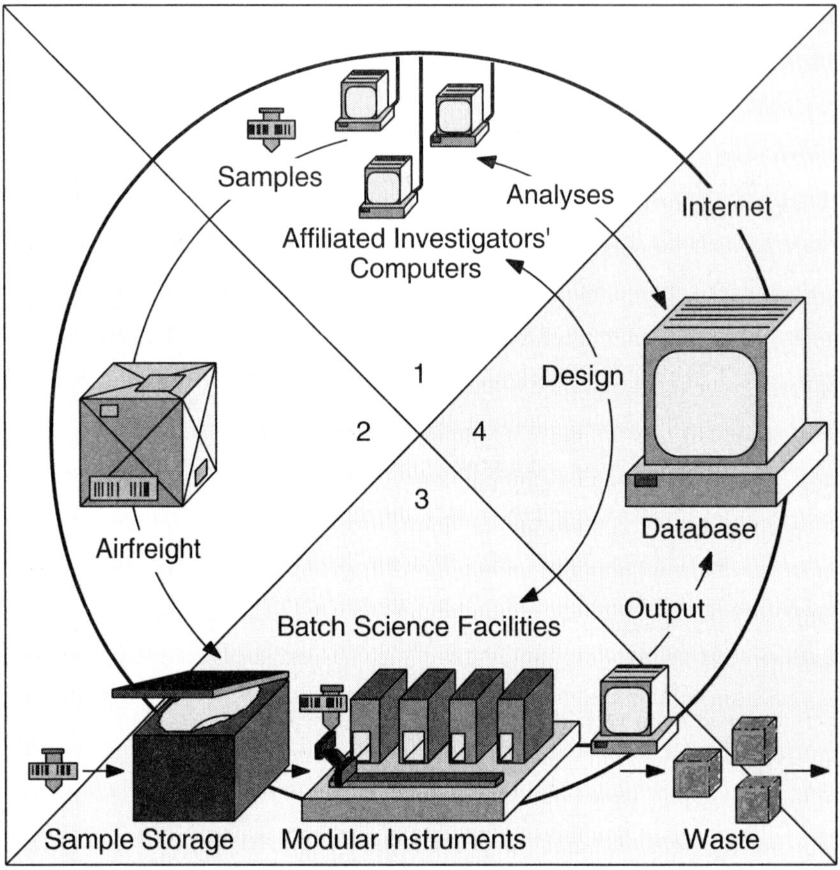

FIGURE 6.2 Batch science via the Internet would enable investigators (at any geographic location) to submit sets of biological samples and automated-instrument instructions as coordinated packets. This approach is distinct from virtual science via the Internet, where experiments are controlled from afar in real time. (1) Scientists located anywhere in the world design experiments in cooperation with a particular batch science laboratory. (2) Shippers deliver packages containing barcoded specimens or reagents. (3) Batch science laboratories house flexible, modular, and scalable instruments. (4) Database facilities maintain permanent records and provide software for analyzing and managing data. The flow of materials and information through the batch operating system is summarized below. The sequence of steps is well suited to most basic science, clinical trials, and public health efforts:

- Investigators download Process Control Tools (PCTs) from the Internet.
- Investigators design assay scripts at home base.

continued

would use this information to address the following key questions as the effort moves forward (Anderson et al., 1997):

- What are the vaccinees' baseline immune parameters?
- What percentage of vaccinees respond to immunization?
- What are the humoral and cellular responses after each dose?
- Does the vaccine demonstrate protection or efficacy?
- How broad are the responses against the collected panel of carrier viruses?
- Are there any significant isolate-specific failures?
- How long does protective immunity persist?

When compared to laboratory procedures performed by human technicians, the quality assurance team would further observe that batch science methodologies are significantly faster, offer larger sample sizes, and exhibit greater reproducibility—all of which would translate into far better clinical trials for AIDS vaccines.

To carry out the various procedures illustrated above, AIDS investigators would use a suite of Process Control Tools (PCTs) to program, coordinate, and track scientific procedures at every step. For instance, PCTs would be used to assign bar codes to samples, to script individualized assay protocols, to analyze raw data, and to create relational links with associated information. Also, at every step along the way private or

---

FIGURE 6.2 Continued

- Investigators collect, package, and bar code numerous samples.
- Investigators document background information.
- Investigators e-mail assay scripts via the Internet.
- Samples and special materials are shipped to the laboratory facility.
- Laboratory receives/logs the bar-coded samples and digital assay scripts.
- Laboratory holds the samples in short-term frozen storage.
- Laboratory pairs the samples with the assay scripts.
- Laboratory provides common supplies and reagents.
- Laboratory performs assays in a batch science mode.
- Laboratory disposes of used specimens and waste materials.
- Laboratory archives certain specimens as ordered by investigators.
- Laboratory forwards raw data to the database.
- Investigators analyze their own data with additional sets of PCTs.
- Investigators link their data to other relevant information.
- Investigators assign data-use privileges.
- The investigator-airfreight-laboratory-database cycle continues.
- Organized digital records amass over time.

BOX 6.2
Major Categories of Assays

AIDS research relies heavily on three major categories of assays. These assays are usually adapted for particular uses. Some of their more common adaptations are listed below.

Assay	*Uses*
Viral infectivity	Drug susceptibility testing (antiviral phenotypes)
	Antibody-blocking activities (humoral immunity)
	HIV isolation (quasi species)
	HIV growth (reproductive numbers)
	Cell tropism (chemokine receptor usage)
	Kinetic measures (spontaneous decay rates)
	Inherent physical properties (infectious fractions)
Molecular sequencing	Drug susceptibility assays (antiviral genotypes)
	Viral load measures (quantitative PCR)
	Classifying subtypes (clades)
	Mutation rates (quasi species versus time)
	Molecular evolution (genotypes versus time)
	Geographic distribution (genotypes versus location)
Cytotoxic T lymphocyte (CTL)	Immune competence (cellular immunity versus time)
	Vaccines (cellular immunity versus immunogens)
	Immunotherapies (cellular immunity versus immunokines)
	Immunogenetics (HLA-restricted epitopes)
	Altered CTL epitopes (escape mutants)
	CTL static killing (percent-specific lysis)
	CTL kinetic killing (viral burst sizes and viral cycle times)

sensitive information would be protected by any number of accepted encryption and authentication methods. This would also be handled by the PCTs (see Box 6.3).

AVAILABLE TECHNOLOGIES

Modular automation and robotics hardware are now available for conducting practically every necessary task in AIDS research—such as bar coding, liquid handling, centrifuging, incubating, sequencing, immunostaining, scintillating, image capturing, and flow cytometer scoring (Garner, 1997). In addition, most major manufacturers of high-throughput laboratory equipment also sell software systems for interconnecting and operating their own products. However, the problem is that such proprietary systems work only so long as customers use equipment from a single company, which is rather unusual in many laboratory settings. Thus, a growing frustration for investigators is that it takes an inordinate amount of time to write new software that integrates components from various manufacturers into one effective machine.

To alleviate such problems, the American Society for Testing and Materials (ASTM) has recently adopted the Laboratory Equipment Control Interface Specification, or LECIS (ASTM, 1998). The new standard will enable manufacturers to create flexible software systems for interconnecting and operating practically any type of compliant hardware, and such innovations will soon lead to high-throughput laboratory systems that exhibit many compatibilities and conveniences found in personal computers. As indicated by the lowermost arrows in Figure 6.3, LECIS defines all the necessary commands between Standard Laboratory Modules (SLMs) and one Task Sequence Controller (TSC). From an investigator's perspective, SLMs are programmable finite-state machines capable of performing various customized tasks in laboratory protocols. Suppose the first few tasks in a viral infectivity assay (for the vaccine trial mentioned above) are to thaw the sample tubes, open the tubes, and then add fresh culture media. In this example, a liquid-handling SLM would perform the following set of tasks under the overall command of the TSC:

SLM ready → input capped/frozen tube → inspect → warm to 37°C →
uncap → add liquid → recap → vortex → verify weight →
output capped/reconstituted tube → SLM ready

In batch science machines, each SLM is dedicated to performing particular parts of laboratory tasks (e.g., moving, measuring, mixing, reacting, culturing) that fall within its physical constraints (e.g., volumes,

BOX 6.3
Process Control Tools (PCTs) to Program, Coordinate, and Track Scientific Procedures.

PCTs would be fashioned to connect investigators from any geographic location to batch science machines and their associated database facilities. PCTs would be designed to permit maximum flexibility and control over experiments by remote investigators, just as though they were employing laboratory technicians to carry out their assay protocols. PCTs would also be designed to interface easily with web browsers (such as the Microsoft Explorer, Netscape Navigator, and Sun Hotjava) and data analysis software packages that are commercially available. To enable just about any type of research activity, one can envision a collection of seven basic PCTs as follows:

- Access PCT would handle the appropriate authorization and gateway functions.
- Operation PCT would describe how to use batch science machines, offer a selection of standardized laboratory tests, and permit investigators to write their own assay scripts.
- Documentation PCT would perform a variety of annotating functions, allowing investigators to deposit relevant background information on specimens, reagents, and related matters. To afford maximum flexibility, the formats could include written, audio, and video documents.
- Submission PCT would tell investigators how to package their specimens and associated reagents and how to generate identifying barcodes to be affixed to all containers before shipping. To support the high-throughput environment, incoming items would be packaged in standardized containers (tubes, bottles) so that they could go directly into batch science machines.
- Analysis PCT would provide parsing tools for manipulating raw data formats, computational tools for analyzing data, and relational tools for linking data to other types of information. It would also provide convenient links to corresponding quality control data from the batch science facilities.
- Privileges PCT would allow submitting investigators to designate who has permission to view or use their data. Feasible options for information management arrangements include (1) access by submitting investigator only, (2) access by certain designated collaborators, (3) time-embargoed data followed by wider access, and (4) unrestricted access by all.
- Commerce PCT would deal with the business aspects of batch science facilities, such as audit trails, billing services, inventory management, and cost modeling.

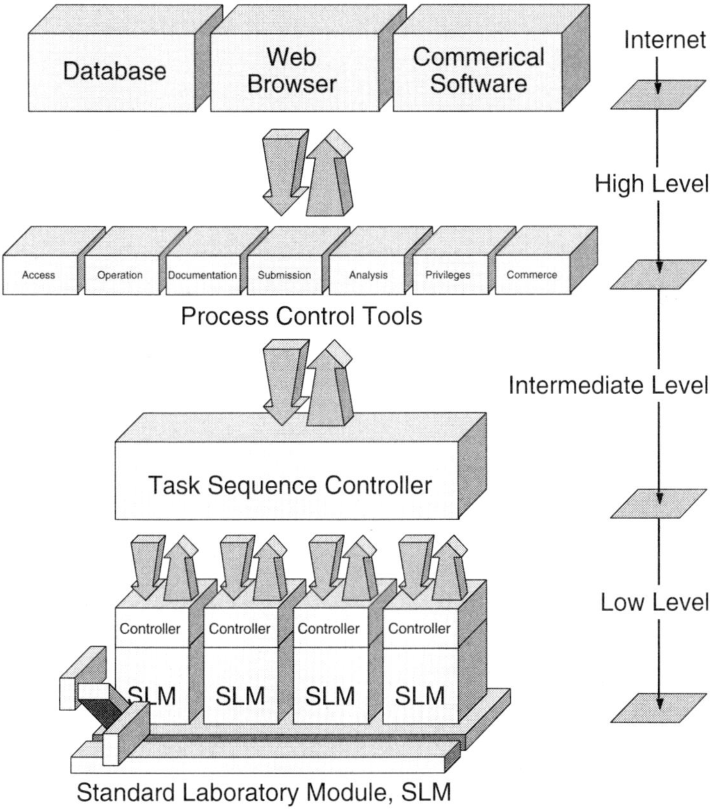

FIGURE 6.3 Batch science machines would operate on a three-tier hierarchy. At the lowest level, modules possess Standard Laboratory Module (SLM) controllers that drive components such as actuators, detectors, and servomotors and coordinate their internal electromechanical activities. Such events are contained wholly within each module (i.e., modules are blind to the existence of one another), which gives rise to independently working SLMs. At the intermediate level, Task Sequence Controllers (TSCs) use tools from operations research to govern intricate flows of supplies and samples through the entire machine. Various standardized commands and feedback signals work to carry out complete laboratory procedures, leading to flexible and programmable assay scripts. At the highest level, PCTs would enable AIDS investigators to carry out a spectrum of important activities via the Internet.

g forces, temperatures) and is also capable of performing these tasks in any arbitrary order (Beugelsdijk, 1989). In theoretical terms there is no upper limit to the number of SLMs per machine or to their physical size. In practical terms we envision that high-throughput machines for AIDS research could be constructed from a collection of merely 10 to 20 different SLMs that cover a full range of tasks. Analogous to computer architecture, duplicate modules can be used to add parallel-processing capabilities to machines, and expansion slots can be used to incorporate new capabilities or technologies. It is also feasible to build batch science machines based on conventional macroscale technologies (e.g., pipettes, test tubes, 96-well plates), newer nanoscale technologies (e.g., microchannels, microchambers, microarrays), or modular combinations of both (Persidis, 1998). Batch science machines thus offer unprecedented capabilities to carry out high-throughput research in a highly flexible operating environment. It now becomes a question primarily of the bigger strategies and directions that AIDS investigators wish to pursue.

THREE ESSENTIAL ASSAYS

Basic science, clinical trials, and public health efforts all rely on a laboratory-based toolbox that consists of viral infectivity, molecular sequencing, and cytotoxic cellular assays (Burton, 1997; Moxon, 1997; Bangham and Phillips, 1997). These three assays are then customized and optimized to meet the particular needs at hand (Box 6.2). Instrument designers would thus consider the most advantageous way to build a small assortment of batch science machines that work together as seamless units (Pollard, 1997). The bottom line is that small clusters of high-throughput machines would be flexible enough to cover the range of tasks that most AIDS investigators demand.

Traditionally, viral infectivity assays have been used in basic research, but more recently they have found clinical utility in the area of drug susceptibility tests (Martinez-Picado and D'Aquila, 1998). Because HIV mutates spontaneously and tolerates any number of amino acid mutations, it can readily escape the inhibitory pressures of reverse transcriptase and protease inhibitors. Consequently, the majority of HIV-infected individuals who take highly active antiretroviral therapies (HAART) are at risk for selecting multidrug-resistant isolates that predominate with time. In clinical practice these complications are first recognized by increasing HIV RNA levels and decreasing $CD4^+$ cell counts, but, if left untreated, they can also proceed to accelerated disease progression (Hecht et al., 1998). Because multidrug-resistant isolates are becoming more common, there are growing demands for drug susceptibility tests (e.g., phenotypic assays) that provide useful information for clinical decision making. However, there are now 11 licensed antiretroviral drugs, and, potentially, each one must be evaluated over a range of concentrations in order to calculate the concentration that inhibits HIV growth by 50 percent (IC_{50}), 90 percent (IC_{90}), or both (Hirsch et al., 1998). The overall complexity and expense of selecting optimal therapies reinforce the need for high-throughput laboratory resources that are highly flexible and that take advantage of economies of scale.

Guiding clinical decisions are just one important use for viral infectivity assays. Others are listed in Box 6.2. Practically any design of infectivity assay would be carried out by three batch science machines that work in series. The first machine would input bar-coded samples, dispense cell cultures (e.g., lymphocytes, monocytes), add customized reagents (e.g., immunoglobulins, antivirals, chemokines), add viral inocula, and output bar-coded plates. The second machine would incubate incoming plates, replenish culture media at specified intervals, and output fully cultured plates. The third machine would score the incubated wells (e.g., by colorimetry, image analysis, flow cytometry) and then send the raw data to the information storage system. From the user's standpoint, all three viral infectivity machines would work together while following the commands of assay scripts (Figure 6.2). In addition, the first and second machines would be capable of performing all the necessary tasks for making recombinant viruses and growing infectious viral stocks.

Molecular sequencing assays have also moved beyond the realm of basic research and into the realms of clinical care and public health efforts. One such application pertains to detecting point mutations in reverse transcriptase and protease enzymes that confer various levels of resistance to one or more antiretroviral drugs. Depending on the choice of laboratory techniques, these genotypic assays are capable of yielding either full-length or partial-length sequences, but, in either instance the amino acid data offer only partial answers (Martinez-Picado and D'Aquila, 1998). In many situations, resistance is conferred not only by the point mutations themselves but also by complex structural and compensatory interactions throughout these viral proteins. To gain more accurate assessments, genotypic assays must therefore be interpreted in conjunction with the phenotypic assays described above (Hecht et al., 1998). This need to perform two categories of laboratory tests and organize multiple sets of data is well suited to batch science methodologies.

Another important application for molecular sequencing assays pertains to identifying the "right" assortment of viral sequences for inclusion as immunogens in future AIDS vaccine trials. One logical approach would consider sequences that represent the most prevalent genotypes from certain geographic regions, but there is no clear consensus on the issue at this time. A comprehensive survey of viral genotypes could certainly help with this identification process; however, such an undertaking will require enormous numbers of epidemiologic samples and sequencing assays (Figure 6.1). AIDS investigators could feasibly conduct such large-scale surveys with flexible PCTs (Box 6.3) that work in conjunction with high-throughput viral sequencing facilities.

Practically any design of molecular sequencing assay would be carried out by three batch science machines. The first would complete sample preparation and polymerase chain reaction (PCR)-amplification steps on cell-free viruses that contain RNA-based cores and cell-associated viruses that harbor DNA-based messages. The second and third machines would work in parallel and house modules for short-length (chip array and mass-spectrometer-based) or long-length (electrophoresis and microcapillary based) genomic determinations, respectively (Garner, 1997). All three molecular sequencing machines would work together in accordance with assay scripts.

In general, cytotoxic T lymphocyte (CTL) assays have been the most difficult to utilize in basic research and clinical management situations. One problem has been the lack of laboratory methods for direct visualization and quantification of antigen-specific cytotoxic T cells (Callan et al., 1998). Another has been the supply of sufficient quantities of HLA-restricted T cells for large-scale research efforts (Schwartz, 1998). But recently a number of innovative laboratory methods have been developed for reducing these problems, enabling AIDS investigators and clinicians to consider how batch science facilities could enhance their work.

Most CTL assays are set up to measure the "percent specific lysis" by chromium release, which is generally scored at one time point after mixing various ratios of effector cells (E) with target cells (T). Although such E:T assays are often used to detect specific CTL activities in vitro, they are limited to providing only static measures of cell killing (Pantaleo et al., 1997; Rosenberg et al., 1997). On the other hand, scoring CTL assays over a series of time points and utilizing new computer-based methods of data analysis would make it possible to provide kinetic measures of cell killing (Spouge and Layne, 1999). In such experiments the control arm would measure HIV reproductive statistics (e.g., viral burst sizes and cycle times) in the absence of effector CTL, while the corresponding experimental arm would measure reproductive statistics in the presence of effector CTL, with all other conditions kept similar to controls. If the cytotoxic cells have specific activity, they would perturb one or more reproductive statistics (e.g., smaller burst sizes, longer cycle times) in a concentration-dependent pattern. If the cytotoxic cells have no specific activity, reproductive statistics would demonstrate only random fluctuations or possibly patterns attributable to nonspecific cellular toxicity.

Intact cellular immunity is now deemed important for survival in

HIV-infected individuals (Pantaleo et al., 1997; Rosenberg et al., 1997). Since cellular immunity is inherently a kinetic process (e.g., cell killing rates versus viral reproduction rates), it is conceivable that time series CTL assays would offer more relevant assessments of cellular immunity than static E:T assays. Nevertheless, CTL assays that are set up for static (one time point) or kinetic (multiple time points) measures of cell killing would be performed by two batch science machines that work in series. The first machine would complete sample preparation and set up the desired CTL assays. The second would score the CTL assays by scintillation or flow cytometry over the desired number of time points, with both working together according to assay scripts.

In many areas of AIDS research it is often difficult to compare the results from one investigator's laboratory with those of another. Different lots of reagents, minor changes in experimental protocols, and inconsistent cell culture conditions can all lead to variable outcomes (Layne et al., 1992). On top of this, viral infectivity, molecular sequencing, and cytotoxic cellular assays display an inherent level of noise—often necessitating many replicate measurements and painstaking quality controls to guarantee reproducibility. Batch science facilities could help AIDS investigators overcome such difficulties in two meaningful ways. First, the programmable machines would have the necessary flexibility and firepower to evaluate which assay formats are the most reliable and reproducible. Such information would enable AIDS investigators to develop an assortment of *standardized assays* that could be offered in the form of digitalized assay scripts. Second, by offering *common resources* to participating investigators (e.g., cell cultures, growth media, viral stocks), batch science facilities would help to eliminate many of the unknowns that limit comparability. The high-throughput environment would permit enormous numbers of experiments to be run in parallel and against a series of standards on a routine basis to maintain reproducibility and reliability.

INTELLECTUAL PROPERTY

With high-throughput technologies changing the means by which research is conducted, it will be important to maintain the traditional reward system for AIDS investigators. At the heart of this system is the freedom to decide how to share data and new information, which can lead to scientific publications and credit for discoveries involving intellectual property (National Research Council, 1997). For each category shown in Box 6.4, data ownership and privileges can be assigned according to the source of financial support.

For the *closed* category, data would belong solely to the commercial organization (such as a pharmaceutical company) that submitted samples

BOX 6.4
Data Ownership and Privileges

Category	*Description*
Closed	Commercial organization pays for research or testing services.
	Organization maintains sole ownership of the data.
Principal investigator	Government agency supports one principal investigator's research.
	Investigator has time-embargoed ownership of the data.
Consortia	Government agency supports multiple principal investigators' research.
	Collaboration has time-embargoed ownership of the data.
Open	Generated by all sources of financial support.
	Anyone can use the data at any time.

and assay scripts and paid for the research or testing services. Upon completing such work, the batch science facility would encrypt and forward all the raw data to the purchasing organization. Afterwards, it would be the organization's responsibility to manage the security of its private property. For a period of time, the facility would also maintain a secure copy of the digital records to assure redundancy and integrity in accordance with contractual agreements.

For the *principal investigator* category, data would belong to the person receiving government grant support for a reasonable period of time—for example, for as long as 2 to 3 years after the grant ends. Good digital practices would be tied to ongoing grant support, requiring each investigator to maintain his or her database records in an orderly manner. After the time embargo had expired, relational links would be attached to the investigator's digital records and the information would become available to others.

For the *consortia* category, data would belong to all of the collaborating investigators for a reasonable period of time, as suggested above. The collaborators would also have responsibility for maintaining their digital records in an orderly manner, most likely under the supervision of the group's database manager. After the time embargo expires, the organized information would become available to others.

For the *open* category, data and its associated links would belong to the public after its quality is assured. The digital records would come from voluntary submissions and time embargoed data that would be released automatically. The main issues would be maintaining backup copies to assure integrity and deciding how to inventory the data and build relational links.

The National Center for Biotechnology Information (*www.ncbi.nlm.nih.gov*), the HIV Sequence Database (*www.hiv.lanl.gov*), the European Molecular Biology Laboratory (*www.embl-heidelberg.de*), and the DNA Data Bank of Japan (*www.ddbj.nig.ac.jp*) are examples of public database facilities in North America, Europe, and Asia, respectively. Some of these organizations have already established guidelines regarding time-embargoed ownership of data, which could provide reasonable starting points for standard agreements pertaining to batch science (Layne and Beugelsdijk, 1998).

IMPLEMENTATION

A blueprint for initiating batch science via the Internet would involve four related areas of effort. *Hardware integration* would deal with building particular batch science machines from commercial hardware, converting commercial modules to SLM-based operation, and constructing machines that work together to serve investigators' needs. *Software integration* would deal with creating a flexible suite of SLM controllers, task sequence controllers, and process control tools that can be used for a variety of AIDS research efforts. *Adapting science* would deal with quality control matters, standardization of assays, and the scaling-up of necessary reagents so that automated instruments can reach their greatest potential. *Laboratory containment* would deal with the design, operation, and maintenance of biohazard facilities dedicated to high-throughput infectious disease research. Of further significance, batch science facilities would reduce occupational exposures to infectious agents and reduce the unit cost of laboratory experiments by taking advantage of economies of scale (see Box 6.5).

AIDS investigators working at global pharmaceutical companies have had strategic access to high-throughput automated laboratory instruments, while investigators working at most major universities have enjoyed far less access to such experimental facilities. This skewed situation, however, is ripe for change. To maintain a pipeline for creating new therapeutics, global pharmaceutical companies are forming increasing numbers of alliances with external academic research groups (Herrling, 1998). Consistent with the trend in research "outsourcing," batch science facilities

BOX 6.5
Batch Science Facilities

Batch science facilities would take advantage of economies of scale, thereby reducing the unit cost of laboratory experiments. Below is an economic comparison of a manual laboratory (employing 100 full-time technicians and occupying ~50,000 sq. ft. of floor space) to a batch science facility (employing 10 full-time technicians and occupying ~5,000 sq. ft. of space) with the same throughput capacity.

Capital costs. Typical biohazard level 3 (BL3) space costs $200 per square foot, and new equipment costs $200 per square foot. In comparison, high-throughput laboratory space costs $400 per square foot, and automation equipment costs $1,000 per square foot. The cost of human labor scales as a linear relationship, whereas the cost of batch science scales as a sublinear one.

Manual Facility	*Cost*	*Batch Science Facility*	*Cost*
50,000 sq. ft. of floor space	$10 million	5,000 sq. ft. of floor space	$2 million
Capital equipment	$10 million	Automation equipment	$5 million
Total	$20 million	Total	$7 million

Operating costs. Typical salary and fringe benefits for a laboratory technician cost $50,000 per year. One technician can perform approximately 400 detailed laboratory procedures annually at an average cost of $250 per procedure, which scales linearly. Automated instruments are more mechani-

they would thus serve as national and international centers for carrying out investigator-initiated research, they would be sponsored by consortia of government agencies and research-driven companies, and would be located at universities or national laboratories. During the start-up period, the center would receive major financial support from sponsoring organizations, permitting the creation and validation of its initial resources. After a center becomes fully operational, the level of start-up support would be scaled down and, over several years would feasibly evolve into a nonprofit cally adept than skilled technicians, and the relative savings could easily amount to a fivefold reduction per test ($50 per procedure) when compared to human-based manipulations.

Manual Facility	*Cost*	*Batch Science Facility*	*Cost*
Salaries for 100 technicians	$5 million	Salaries for 10 technicians	$0.5 million
40,000 assays	$10 million	40,000 assays	$2 million
Maintenance	$2 million	Maintenance and upgrades	$2 million
Yearly total	$17 million	Yearly total	$4.5 million

Five-year average. Over 5 years both facilities would perform a total of 200,000 major laboratory procedures. Batch science facilities would reduce the unit cost of laboratory testing by over threefold. The overall savings would make it possible for investigators to carry out larger experimental undertakings for the same dollar expenditures.

Manual Facility	*Cost*	*Batch Science Facility*	*Cost*
Initial capital costs	$20 million	Initial capital costs	$7 million
Five-year operating costs	$85 million	Five-year operating costs	$22.5 million
Five-year total	$105 million	Five-year total	$29.5 million
Unit cost per assay	~ $525	Unit cost per assay	~ $150

entity that charges investigators reasonable fees for "mass-customized" testing services.

All too often, AIDS investigators shy away from problems that deal with the bigger picture simply because they lack the necessary tools to move ahead. In HIV/AIDS research, there are many important problems to be solved that seem beyond reach because of the enormous "experimental spaces" to be explored and characterized (Fields, 1994). The HIV-RNA genome is composed of ~10^4 bases, which corresponds to ~3,000 amino acids. If only a small fraction (~1 percent) of amino acids were responsible for conferring unique viral properties, and only one specific amino acid (from the total selection of 20) conferred this property, there could be upwards of $2^{30} \approx 10^9$ unique variations (see Figure 6.4). Even if only one in a thousand mutations produced viable offspring, there would still be $\sim 10^6$ variants, which is an enormous number to attempt to sample and characterize. At present, we have only sketchy information mapping genotypic variations to phenotypic expressions, making it difficult to relate the significance of such variations (if any) to therapies and vaccine development (Figure 6.1). And what little knowledge we do have is hopelessly scattered among various investigators' notebooks. Batch science facilities, with the capacity to perform the work of hundreds of laboratory technicians, would enable investigators to conceptualize and attack problems from entirely new directions. For many such problems with large *phase spaces*, high-throughput laboratory facilities with digitalized record-keeping are perhaps the only feasible means for moving ahead.

Health care is progressing in the direction of therapies that are precisely tailored to individuals or smaller groups of patients. In this regard, HIV-infected individuals who take antiviral drugs must rely on precisely the "right" combination (e.g., reverse transcriptase plus protease inhibitors) to halt viral replication (Deeks and Abrams, 1997). With current clinical and laboratory practices, however, there is no single test to establish the optimal regimen. Therefore, to maintain effective therapies, physicians have come to rely on a series of tests (i.e., viral loads, molecular sequencing assays, drug susceptibility assays) that are highly repetitive, time consuming, and labor intensive. Batch science facilities would assist by providing the necessary high-throughput testing services and creating an organized base of information for clinical decision making. In addition, pharmaceutical companies could use batch science facilities to track the molecular evolution of HIV in patients around the world. A wider base of information could then be used to guide the development of new generations of antiviral drugs.

In the Third World, dedicated AIDS investigators are often forced to work with primitive laboratory resources. For such infectious disease research, however, this is precisely where any number of meaningful studies should be conducted. With simple resources in the field—an Internet connection, plasticware, barcodes, refrigeration, and air transportation—one batch science facility could serve large geographical regions.

The creation of batch science facilities will not require a Manhattan Project (nor major redirections of resources) as some might suspect. On the contrary, it will involve a focused effort by a relatively small team of motivated engineers and scientists. With appropriate support, the first-generation facilities could be running within 2 to 3 years, and next-

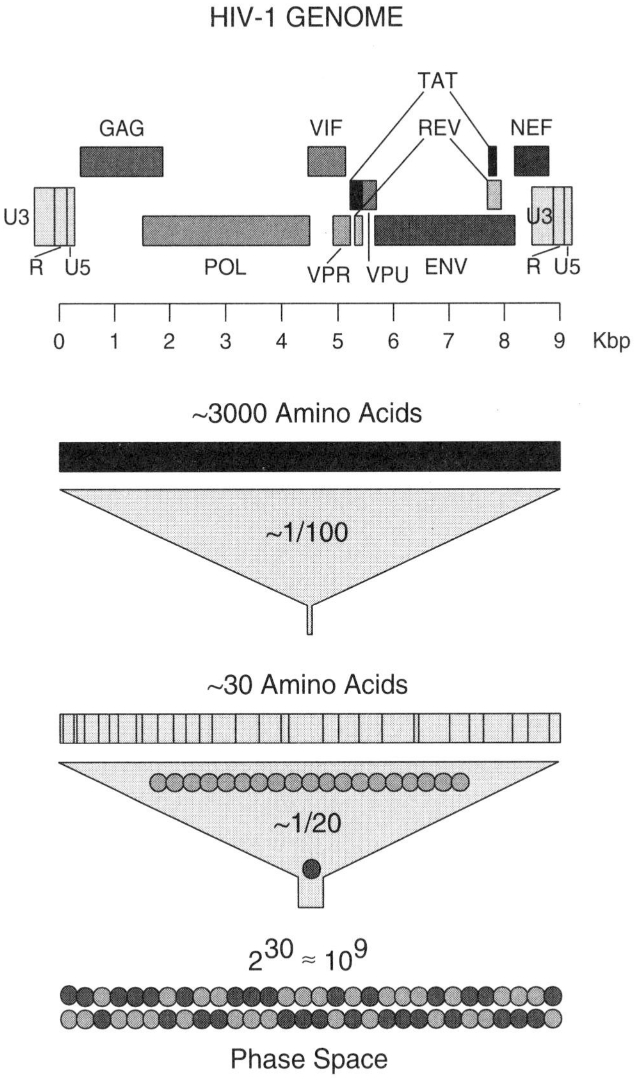

FIGURE 6.4 Estimating the number of HIV-1 isolates by combinatoric arguments. The entire HIV-RNA genome codes for ~3,000 amino acids and mutations in many of these positions may lead to viable viruses. If we conservatively assume that only ~30 of these positions (~1/100) are responsible for conferring phenotypic properties and that only one amino acid out of a pool of 20 (~1/20) is capable of conferring viable viruses, there could be upward of $2^{30} \approx 10^9$ viable genotypes and unique phenotypes. There are many ways to pose such combinatoric arguments, but they all generate large numbers.

generation facilities would soon follow. If we assume that the facilities perform the work of, say 100 laboratory technicians at a fraction of the cost, the development costs would be recouped within several years. The time is right for government funding agencies, the World Health Organization, health-related foundations, and global pharmaceutical companies to consider their respective roles in building "intermediate-scale" research infrastructures that are comparable to those in the Human Genome Project. AIDS investigators working in basic science, clinical trials, and public health efforts will have no problems in finding clever (and unforeseen) ways to use flexible laboratory firepower.

REFERENCES

American Society for Testing and Materials (ASTM), Subcommittee E-49.52. 1998. Laboratory Equipment Control Interface Specification (*www.thermal.esa.lanl.gov/astm*).

Anderson, R. M., C. A. Donnelly, and S. Gupta. 1997. Vaccine design, evaluation, and community-based use for antigenically variable infectious agents. Lancet, 350:1466-1470.

Baggiolini, M. 1998. Chemokines and leukocyte traffic. Nature, 392:565-568.

Bangham, C. R. M., and R. E. Phillips. 1997. What is required of an HIV vaccine? Lancet, 350:1617-1621.

Beugelsdijk, T. J. 1989. Trends for the laboratory of tomorrow. Journal of Laboratory Robotics Automation, 1:11-15.

Burton, D. R. 1997. A vaccine for HIV type 1: The antibody prospecting. Proceedings of the National Academy of Sciences USA, 94:10018-10023.

Callan, M. F., L. Tan, N. Annels, G. S. Ogg, J. D. Wilson, C. A. O'Callaghan, N. Steven, A. J. McMichael, and A. B. Rickinson. 1998. Direct visualization of antigen-specific $CD8^+$ T cells during the primary immune response to Epstein-Barr virus in vivo. Journal of Experimental Medicine, 187:1395-1402.

Chumbley, L. S., M. Meyer, K. Fredrickson, and F. Laabs. 1995. Computer networked scanning electron microscope for teaching, research, and industrial applications. Microscopy Research Technique, 32:330-336.

Deeks, S. G., and D. I. Abrams. 1997. Genotypic-resistance arrays and anti-retroviral therapy. Lancet, 349:1489-1490.

Fields, B. N. 1994. AIDS: Time to turn to basic science. Nature, 369:95-96.

Garner, H. R. 1997. Custom hardware and software for genome center operations: From robotic control to databases. Pp. 3-41 in Automation Technologies for Genome Characterization, T. J. Beugelsdijk, ed. New York: John Wiley.

Gazzard, B. G., and G. J. Moyle, on behalf of the BHIVA Guidelines Writing Committee. 1998. 1998 revision to the British HIV Association guidelines for antiretroviral treatment of HIV seropositive individuals. Lancet, 352:314-316.

Germain, E. 1996. Fast lanes on the Internet. Science, 273:585-588.

Hecht, F. M., R. M. Grant, C. J. Petropoulos, B. Dillon, M. A. Chesney, H. Tian, N. S. Hellmann, N. I. Bandrapalli, L. Digilio, B. Branson, J. O. Kahn, et al. 1998. Sexual transmission of an HIV-1 variant resistant to multiple reverse-transcriptase and protease inhibitors. New England Journal of Medicine, 339:307-311.

Heilman, C. A., and D. Baltimore. 1998. HIV vaccines—Where are we going? Nature Medicine (Suppl.), 4:532-534.

Herrling, P. L. 1998. Maximizing pharmaceutical research by collaboration. Nature (Suppl.), 392:32-35.

Hirsch, M. S., B. Conway, R. T. D'Aquila, V. A. Johnson, F. Brun-Vezinet, B. Clotet, L. M. Demeter, S. M. Hammer, D. M. Jacobsen, D. R. Kuritzkes, C. Loveday, J. W. Mellors, S. Vella, and D. D. Richman. 1998. Antiretroviral drug resistance testing in adults with HIV infection: Implications for clinical management. International AIDS Society—USA Panel. Journal of the American Medical Association, 279:1984-1991.

Lanier, B. M., V. G. Cerf, and D. D. Clark. 1997. The past and future history of the Internet. Communications ACM, 40:102-108.

Layne, S. P., and T. J. Beugelsdijk. 1998. Laboratory firepower for infectious disease research. Nature Biotechnology, 16:825-829.

Layne, S. P., M. J. Merges, M. Dembo, J. L. Spouge, S. R. Conley, J. P. Moore, J. L. Raina, H. Renz, H. R. Gelderblom, and P. L. Nara. 1992. Factors underlying spontaneous inactivation and susceptibility to neutralization of human immunodeficiency virus. Virology, 189:695-714.

Martinez-Picado, J., and R. D'Aquila. 1998. HIV-1 drug resistance assays in clinical management. AIDS Clinical Care, 10:81-88.

Moxon, R. E. 1997. Applications of molecular microbiology to vaccinology. Lancet, 350:1240-1244.

Murray, C. J. L., and A. D. Lopez. 1997. Alternative projections of mortality and disability by cause 1990-2020: Global Burden of Disease Study. Lancet, 349:1498-1504.

National Research Council, Computer Science and Telecommunications Board. 1993. National Collaboratories. Washington, D.C.: National Academy Press.

National Research Council, Committee on Intellectual Property Rights and Research Tools in Molecular Biology. 1997. Intellectual Property Rights and Research Tools in Molecular Biology. Washington, D.C.: National Academy Press.

Pantaleo, G., H. Soudeyns, J. F. Demarest, M. Vaccarezza, C. Graziosi, S. Paolucci, M. Daucher, O. J. Cohen, F. Denis, W. E. Biddison, R. P. Sekaly, and A. S. Fauci. 1997. Evidence for rapid disappearance of initially expanded HIV-specific $CD8^+$ cell clones during primary HIV infection. Proceedings of the National Academy of Sciences USA, 94:9848-9853.

Persidis, A. 1998. High-throughput screening. Nature Biotechnology, 16:488-489.

Pollard, M. J. 1997. Automation strategies: A modular approach. Pp. 43-63 in Automation Technologies for Genome Characterization, T. J. Beugelsdijk, ed. New York: John Wiley.

Rosenberg, E. S., J. M. Billingsley, A. M. Caliendo, S. L. Boswell, P. E. Sax, S. A. Kalams, and B. D. Walker. 1997. Vigorous HIV-1–specific $CD4^+$ T cell responses associated with control of viremia. Science, 278:1447-1450.

Schwartz, R. S. 1998. Direct visualization of antigen-specific cytotoxic T cells—A new insight into immune disease. New England Journal of Medicine, 339:1076-1078.

Spouge, J. L., and S. P. Layne. 1999. A practical method for simultaneously determining the effective burst sizes and cycle times of viruses. Proceedings of the National Academy of Sciences USA, 96:7017-7022.

7

Input/Output of High-Throughput Biology: Experience of the National Center for Biotechnology Information

David J. Lipman

INTRODUCTION

This paper describes several current projects being conducted by the National Center for Biotechnology Information (NCBI), a division of the National Library of Medicine, that have involved collaboration among numerous groups. It will also review some aspects of how these projects were organized and developed.

MEDLINE/PUBMED

The PubMed system provides access to Medline, a database of more than 10 million abstracts in the biomedical literature, and also features links to online journals and a number of factual databases.[1] PubMed currently receives approximately 8 million hits each day. Although the development of PubMed may not appear to involve high-throughput biology, this system does illustrate how many biology databases were built in the past, as well as their strengths and weaknesses.

First, the authors of the articles that are accessed through PubMed do not participate in its development and are not required to have knowledge of the database's functionality. The information is mainly keyed in from the journals, although currently there is some use of electronic input

---

[1]PubMed, Medline, and MeSH are registered trademarks of the National Library of Medicine.

and optical character recognition. Nearly 400 journals submit electronic header information, which is collected in a dataset called Pre-Medline.

It is also important to note that an extensive amount of manual work may be necessary when different groups produce the data and build the database. Medline, for example, requires the assignment of keywords, a process that involves many indexers who read the entire article and then apply appropriate terms from a controlled vocabulary. In fact, a limiting factor for Medline, in terms of the number of journals included and the speed at which the records can be completed, is the cost of this indexing.

Both the availability and the quality of information are important factors for certain databases. For example, with PubMed a search can be conducted using the title and abstract alone, and this information can be made available as soon as the publisher provides it—even before the minimal checking performed at this stage is completed. Other databases also use this process, one that may be particularly relevant for rapid dissemination of information in the area of infectious diseases.

GENBANK

GenBank,[2] a database of gene sequences, now contains over 4 billion base pairs of DNA from more than 60,000 different species (Benson et al., 2000). Unlike most factual databases, GenBank is used for computation. In fact, most of the new biological databases being developed are designed for computation, which requires information to be validated in a variety of ways. For example, sufficient similarity can often be found between the protein from a human gene and one from another organism (e.g., yeast) such that the related yeast sequence can be retrieved through a database search. The information that is already known about the yeast sequence in experiments can then be used. Even if it is not known what the sequence or the protein does in yeast, experiments can be conducted in yeast that cannot be conducted in humans. This is what makes comparative sequence analysis such a powerful tool.

At GenBank, considerable work goes into building the database, using various tools that have been developed for data computation. In addition, a number of tests are conducted at different steps along the way to determine whether the genes are being translated to the protein correctly. The assumption is that a number of the databases that would be developed for high-throughput infectious disease projects would require similar data validation.

[2]GenBank is a registered trademark of the U.S. Department of Health and Human Services.

The data flows for Medline and GenBank differ in a number of ways. For Medline, processing cannot begin in terms of keyboarding or MeSH (Medical Subject Headings) indexing until the journal arrives in its hard-copy format. For GenBank all of the data are electronic, and a submission program is used to help ensure that the syntax of the GenBank submission is preserved. However, most authors provide only annual submissions to GenBank, primarily because they have other priorities, and are unlikely to devote time to understanding the syntax and semantics of the records with which they are dealing. Thus, a disadvantage of this system is that, although authors can produce syntactically satisfactory records, staff skilled in molecular biology must review the biological content.

It is also important to keep in mind that while authors use sequences to answer specific questions, others who will be using GenBank may not be trying to answer those particular questions. In addition, because these authors were only interested in answering specific questions, they may have failed to provide information that, although available, did not relate to those more general questions. This is a common challenge encountered in developing any database: Those who are providing the data may have one main purpose in mind, while the database developer has a much broader range of purposes.

In the past, GenBank used indexers who manually scanned journals to locate sequence data in the articles. Although it would seem that direct electronic submission would be more efficient, in some ways this is not true. While obtaining the sequence electronically is certainly valuable, often authors provide additional information in their papers, and one might more easily find the information needed by reading the paper than by communicating with the author via e-mail or by telephone or fax.

Thus, although much progress has been made, challenges remain, and the process continues to be expensive and relatively slow. For example, annotating the records continues to require an extensive amount of time, notwithstanding the fact that GenBank annotators are highly trained and educated. In addition, senior scientists on NCBI staff are on call to conduct final reviews of these records.

Over time the genome centers will be providing a huge volume of data, and NCBI can work with the centers to make sure they can develop a mode of submission with the correct syntax and semantics. Because in most cases these centers are funded for an infrastructure project—to actually do the sequence rather than answer a particular question—they are very cooperative. NCBI fully appreciates this cooperation and recognizes that staff at the genome centers are vastly overworked. However, it is often difficult to obtain the information that is needed from these centers. This is because many staff members have had long experience with this kind of sequencing and still believe that once they post the data on their website, the job is done.

In planning any high-throughput molecular epidemiology project, it is crucial to ensure that, from the beginning, the groups that are developing and generating the data understand that cooperation with those who are collecting the data makes a significant difference in how the information can be ultimately used.

Multiple collection points exist for all of these kinds of data. With two other databases of DNA sequences, the DNA Databank of Japan (DDBJ) and the EMBL Library of Europe, there is general agreement on underlying syntax and semantics, and daily exchanges occur among these centers, with some high-throughput genomes sent to one group and some to another. The level of agreement here is fairly basic, and at times a genome center has information that one of the collection databases cannot represent in its internal data structures. That information is sometimes lost, and this results in complicated arrangements in which, for example, a center in the United Kingdom submits its primary information to EMBL but also submits some additional information directly to NCBI.

Finally, GenBank provides data in a number of ways. It is available on NCBI's site for interactive use or for computing via the BLAST (Basic Local Alignment Search Tool) program (Altschul et al., 1990; 1997). It is also available via FTP to be downloaded and integrated with other datasets. With any of these projects that are creating an information infrastructure in a particular area, it is important to make the data as widely available as possible. It is more effective for a variety of companies and other academic sites to work to make the data available than it is to rely on one center to provide access to the data.

Medline and GenBank also share a characteristic that makes them a bit more challenging to build than some other databases: they are continually being updated, which makes information handling much more complex. Another project, the transcript map, provides periodic releases that involve complete recomputations, an approach that makes information handling much easier. Thus, with a high-throughput project a full re-release every few months makes information handling a less expensive and more efficient process and also contributes to the development of a much more robust system than one in which the information is being updated continually.

HUMAN GENOME MAP

A very different project from the information-handling point of view, and an extremely interesting model, is the human gene map. There have been two major releases of this project—one in 1996 (Schuler et al., 1996) and one in 1998 (Deloukas et al., 1998), which involved an international mapping consortium of genome centers in the United States, the United Kingdom, France, and Japan. Approximately 16,000 genes were mapped in the first release, and then more than 30,000 genes were mapped in the second release. This is very useful for gene hunting. If a region of the genome can be identified based on affected families that may have a disease gene, one can then click on that little part of the map or input the two markers and find all of the human genes that have already been mapped there. A region can be input to get all of the markers that are in that region, or a gene itself can actually be entered. In some cases a database search can make a year's difference in terms of discovering the gene.

The evolution and information-handling aspects of this project are interesting. It began in an informal manner, with no directly targeted funding. Several mapping groups and Greg Schuler at NCBI developed a way to build the map, and ultimately the first release was so successful that additional funding was received to continue the project. One of the consequences of developing the project in this way was that a tremendous amount of cooperation emerged among all of the participating groups. From the beginning, all of the parties involved functioned as one team, something that is not always possible.

The process of developing the gene map involves clustering partial gene sequences called expressed sequence tags (ESTs). NCBI conducted the computational clustering task and developed a resource called UniGene (Schuler, 1997), which, while initiated specifically for the gene map project, has received far greater use. ESTs are clustered, and then nonredundant representatives are selected from each gene; a sequence of that EST is sent to one of the genome centers. This is an *information* transfer process; a reagent is not actually being sent. The genome centers take these sequences and, using the PCR technique, develop unique markers for the ESTs, which are mapped on two different radiation hybrid panels. Several laboratories are involved at any given time developing many of these markers, which are then sent to a database at the European Bioinformatics Institute called RHdb (Radiation Hybrid Database). Next, two centers download the data from RHdb and construct a map using that information. Map construction involves complete recomputation each time it is revised.

Through an ongoing cycle of checking data and relaying information back to the centers, NCBI can refine the quality of the data and improve the quality of its own clustering. The feedback provided is also useful to improve the quality of intermediate information resources, which are useful internally and also distributed via the World Wide Web.

CANCER GENOME ANATOMY PROJECT

The Cancer Genome Anatomy Project (CGAP) provides a model for future full-length cDNA projects by the National Institutes of Health (NIH). Although NCBI is involved in this effort, CGAP is funded and directed by the National Cancer Institute (NCI). The project has developed information about reagents for deciphering the molecular anatomy of the cancer cell. From the beginning, this project developed not just information but reagents as well. A great deal of information and many ways of querying the system are available on this site. One of the main goals is to develop a tumor gene index of all of the genes that are involved in cancer. Obviously, because any gene could have something to do with cancer, in some sense this involves trying to find all the human genes. However, the goal is to find the genes that are the most likely to be involved in cancer. which is not always consonant with the other main goal of this project—to maximize the gene discovery rate. Achieving a balance between these two goals presented a major challenge during the early part of this project.

Tissues and cell lines are obtained from a variety of sources, including normal colons, precancerous colons, and cancerous colons; cDNA libraries are made; and ESTs are sequenced. Then, a variety of analyses are performed, and that information and the reagents, the actual clones, become one of the products. The overall gene discovery rate is updated weekly. Because a tailing off of the discovery rate was evident with the previous EST project, a great deal of effort went into monitoring progress and maintaining as high a discovery rate as possible.

It appears that it is best to sequence a library broadly and then sequence it more deeply if it appears to be promising. The data flow is complicated. A steering committee for CGAP has decision-making and spending authority, with most funds spent outside NIH on various contracts. NCBI, a member of the steering committee, has built the tracking database and has modified two existing databases, UniGene and dbEST, to provide reports.

NCI handles the step that involves acquiring the tissues, which come from many different sources, and maintains a repository of these sources when possible. The tissue is then sent to the various groups that build the cDNA libraries. Almost all of this occurs outside NIH. Information is fed into the database and then goes to the cDNA library groups, which send information to the tracking database. The library then goes to a group at Lawrence Livermore Laboratory, part of the consortium, where the clones are arrayed out of the library and reports are sent back to the tracking database. The arrays are then sent to the Washington University group for EST sequencing and to commercial distributors. Tracking is crucial, as it will be important to identify the source of the clones in the database.

All of the reports that NCBI provides are available to the various partners for use in semi autonomously adjusting their work. But sometimes at this level users find problems with the clones, including diverse artifacts and errors. Because NCBI does not have control over the system, these problems are difficult to correct.

In CGAP the various commercial distributors are independently trying to make their own curated, cleaned-up sets of the clones, which can cause confusion in terms of defining the "standard" clone set. Thus, on a new and related initiative—the full-length cDNA project—a steering committee will conduct some of the work itself, and a master repository located at NIH will be responsible for quality assurance of the reagents.

SUMMARY

High-throughput biology projects are representative of database projects of the future, which will involve ever-increasing numbers of participating sites. The challenge is and will continue to be maintaining control and organization in the data collection process while at the same time preserving the system flexibility and autonomy that are essential to the performance of high-quality scientific research.

REFERENCES

Altschul, S. F., W. Gish, W. Miller, E. W. Myers, and D. J. Lipman. 1990. Basic local alignment search tool. Journal of Molecular Biology, 215:403-410.

Altschul, S. F., T. L. Madden, A. A. Schaffer, J. Zhang, Z. Zhang, W. Miller, and D. J. Lipman. 1997. Gapped BLAST and PSI-BLAST: A new generation of protein database search programs. Nucleic Acids Research, 25:3389-3402.

Benson, D. A., I. Karsch-Mizrachi, D. J. Lipman, J. Ostell, B. A. Rapp, and D. L. Wheeler. 2000. GenBank. Nucleic Acids Research, 28:15-18.

Deloukas, P., G. D. Schuler, G. Gyapay, E. M. Beasley, C. Soderlund, P. Rodriguez-Tome, L. Hui, T. C. Matise, K. B. McKusick, J. S. Beckmann, et al. 1998. A physical map of 30,000 genes. Science, 282:744-746.

Schuler, G. D. 1997. Pieces of the puzzle: Expressed sequence tags and the catalog of human genes. Journal of Molecular Medicine, 75:694-698.

Schuler, G. D., M. S. Boguski, E. A. Stewart, L. D. Stein, G. Gyapay, K. Rice, R. E. White, P. Rodriguez-Tome, A. Aggarwal, E. Bajorek, et al. 1996. A gene map of the human genome. Science, 274:540-546.

8

Applications of Modern Technology to Emerging Viral Infections and Vaccine Development

Gary J. Nabel

INTRODUCTION

The rapid spread of HIV as an emerging virus serves as a reminder of the continued susceptibility of large segments of the world population to infectious agents. HIV has adopted several different strategies by which to evade immune detection that have promoted its spread (Bagarazzi et al., 1998; Burton and Moore, 1998; Heilman and Baltimore, 1998). First, by attacking the $CD4^+$ T cell, a major host cell that defends against foreign pathogens, the virus eliminates the specific cell type perhaps best suited for immune protection. Second, the virus encodes gene products that alter the immune recognition and activation. These viral proteins include *Nef*, which can decrease the expression of both CD4 and class I MHC molecules and participate in cell-cell interactions that lead to the detection of foreign antigen. Third, the virus undergoes rapid genetic mutation and has been selected to tolerate a number of changes at the nucleotide level and is therefore a "moving target" for immune recognition. Finally, persistent viral infection in HIV-seropositive individuals increases the duration of infection and the likelihood of transmission, particularly during its asymptomatic phase.

For these reasons, HIV has spread throughout the world, causing infection in more than 30 million people (Palella, 1998). Current projections estimate that an additional 5.8 million individuals are likely to become infected within the next year (UNAIDS, 1998). In addition, the virus has developed heterogeneity in different parts of the world, and it appears that specific forms predominate in certain populations. For example, clade B is most common in the United States, whereas clades C and E are becoming prevalent in East Asia and Africa (Gao et al., 1996a). The development of new clades and continual genetic variability provide the basis for further genetic drift, migration, and infection of new populations. In some cases the characteristics of this virus alter its mode of spread; for example, there is some evidence for differences in sexual transmission patterns from clade to clade (Gao et al., 1996b). In addition to clade differences among HIV strains, two different HIV strains—HIV-1 and HIV-2—provide for further genetic diversity and potential recombination between related regions of these viruses. Thus, it is critical that appropriate detection and surveillance are maintained in order to understand the patterns of spread of different viral types. Effective surveillance will require appropriate cost-effective, high-throughput technology, which has not been fully developed. This information may be used to predict the future populations at risk and the potential for altered pathogenicity in these populations.

EBOLA VIRUS AS ANOTHER MODEL OF EMERGING INFECTION

HIV provides an example relevant to the emergence of other viruses in the world population. Ebola virus represents another recently emerged virus responsible for several outbreaks in Africa, Europe, and North America (see Figure 8.1). These outbreaks have often been associated with infected primates, followed by person-to-person spread. Of great concern regarding Ebola virus infection is its high mortality rate, in some cases up to 90 percent. Several aspects of its biology raise these concerns. Although the virus is readily contained using barrier precautions, the reservoir for infection is currently unknown. The virus is poorly recognized by the immune system and is said to be nonimmunogenic, with no vaccines or treatments for it currently available. New technologies may soon make it possible to rapidly implement vaccination approaches for diseases such as Ebola and HIV. While HIV has proven to be a difficult target for vaccine development, preliminary studies with Ebola virus appear to be more encouraging. Improved technology, however, could help to contain the spread of both of these viruses.

In the case of Ebola virus, at the National Institutes of Health (NIH) we have used newer techniques involving DNA immunization to control infection and have found that the use of plasmid expression vectors that encode Ebola virus gene products can be used for successful genetic immunization in rodent models. In a guinea pig model that resembles human disease, immunization induces both humoral immunity and cellular immunity that confer protection against lethal challenge by the

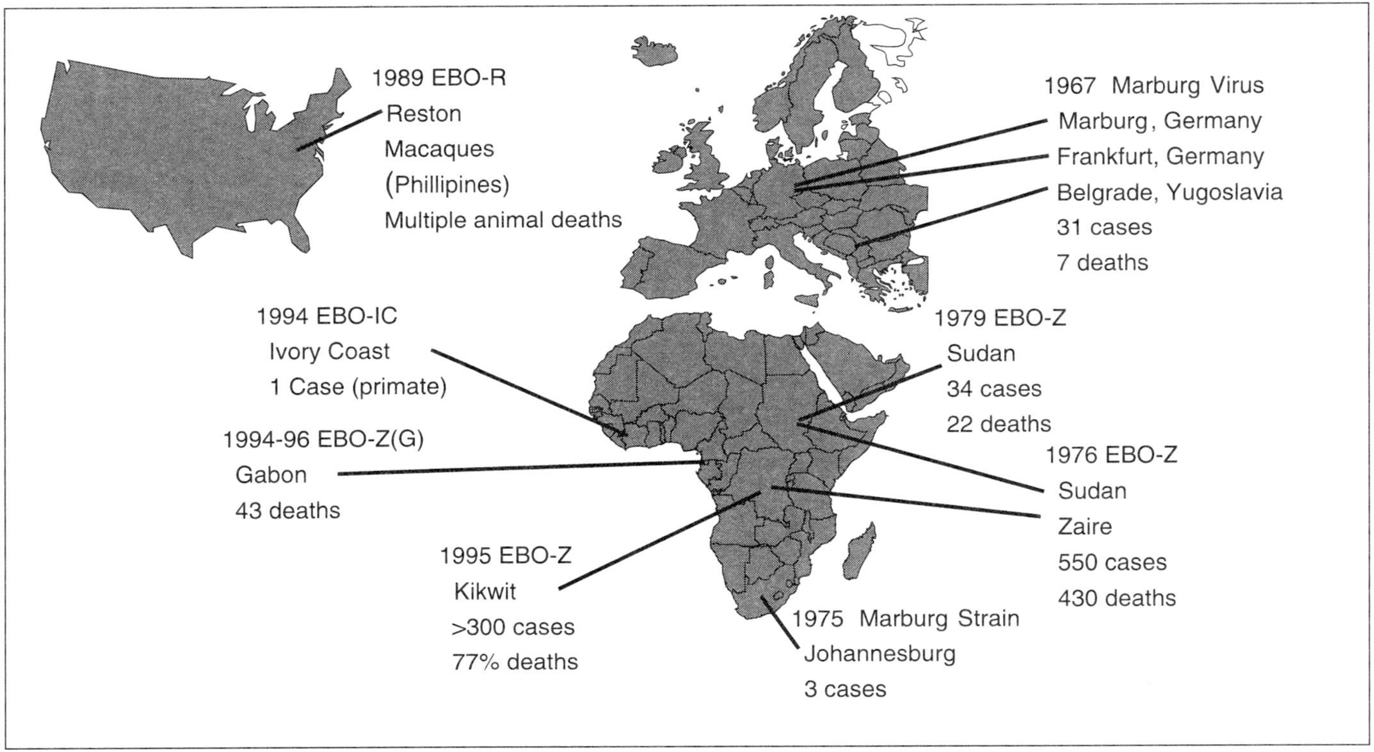

FIGURE 8.1 Chronology and geographic distribution of Ebola virus outbreaks.

virus (Xu et al., 1998). It appears that T-cell-immunity is required for this protection, as measured by an indirect immune parameter, the antibody response to the viral glycoprotein (GP), a T-cell dependent response. Animals that have titers greater than 1:5,000 are nearly completely protected when exposed to a lethal dose of the virus that otherwise causes mortality within a week in unprotected recipients (see Table 8.1). Recently, this approach has been applied, with an adenoviral vector boost, to confer protective immunity against lethal Ebola virus challenge in nonhuman primates (Sullivan et al., 2000).

These approaches are being applied by many laboratories to HIV and to a number of other infectious agents, including tuberculosis, influenza virus, malaria, and herpesvirus (Tighe et al., 1998; Donnelly et al., 1997). Despite promise for many pathogens, difficulties remain in generating successful vaccines for HIV. Such limitations point to a number of stumbling blocks for vaccine development, an area in which bioinformatics and genetic analyses may provide new opportunities and prove useful.

An illustration of the complexity and unpredictability of immune responses to foreign proteins comes from our studies of the immune response to different viral glycoproteins (see Figure 8.2). In the case of Ebola virus as well as HIV, specific viral genes induce characteristic immune responses that are specific to particular proteins. For example, DNA immunization with expression vectors encoding gp160 readily induces a cytolytic T-cell response to this protein but fails to induce a robust antibody response (data not shown). In contrast, vaccination to another HIV gene product, *Nef*, which plays an important role in increasing the pathogenicity of infection and the onset of AIDS symptoms, leads to the generation of excellent humoral immunity but poor cytolytic T-cell responses. Analogous observations have been made in other systems. For example, the Ebola virus glycoprotein readily induces cytolytic T-cell responses but does not induce high-titer antibody responses. In contrast, the nucleoprotein can induce titers greater than 1:60,000 in mice but stimu-

TABLE 8.1 Protection Against Lethal Challenge Correlates with Antibody Titer to Viral Glycoprotein (GP)

Antibody titer (gp)	Survival
> 1:5,120	11/11*
> 1:2,560	13/19**
Control (undetectable)	0/10

*$p = 10^{-4}$
** $p = .00043$

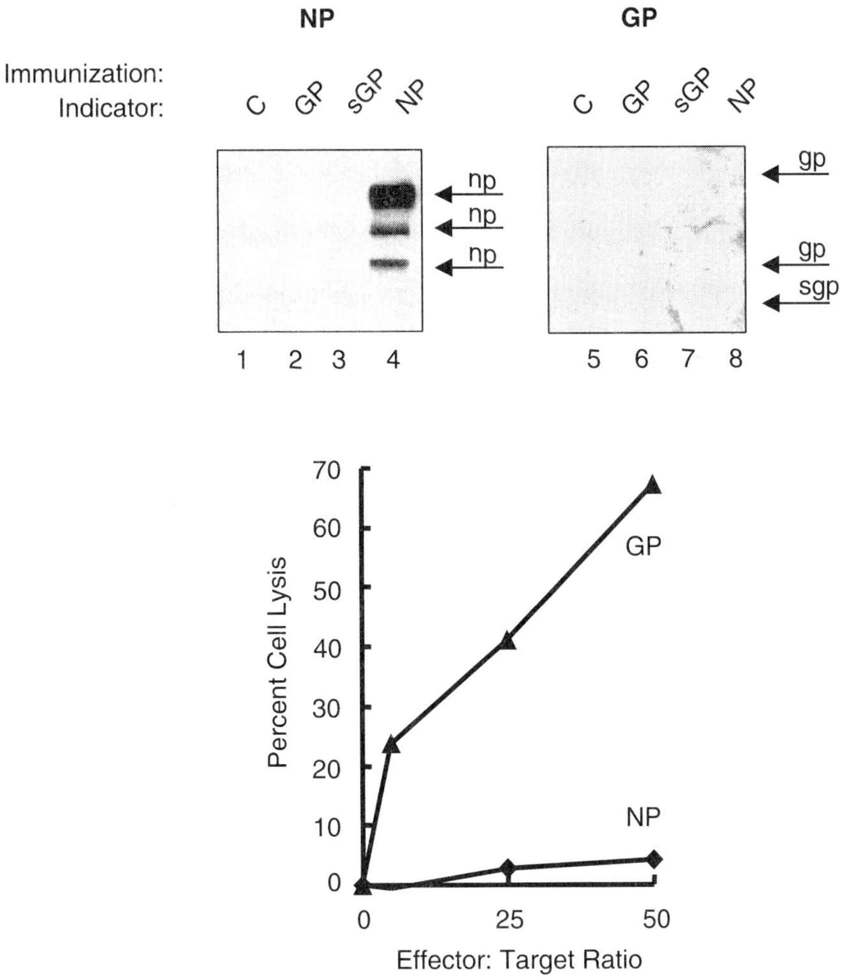

FIGURE 8.2 Inherent differences in intrinsic immune responses to Ebola gene products.

lates essentially no cytolytic T-cell response (Xu et al., 1998). The nature of these immune responses is not random. In fact, it is likely that these viruses have evolved such gene products not only to facilitate viral replication but also to evade immune detection. In Ebola virus infection, the virus displays high replication potential and is likely to rely on innate immune and inflammatory responses to limit spread (Baize et al., 1999; Nabel,

1999), whereas lentiviruses cause a more indolent disease course, with immune mechanisms playing a more important role.

ROLE OF INFORMATION TECHNOLOGIES IN VACCINE STRATEGIES

The challenge is to use information technologies to understand the mechanisms by which specific viral genes induce their characteristic immune responses by using different vaccine strategies and to develop algorithms by which these immune responses can be predicted. The development of approaches by which specific peptides can be made more immunogenic in a predictable fashion using alternative vectors or adjuvants in humans will facilitate the development of effective vaccines and lead to more rational and efficient vaccine approaches. HIV gp160 and *Nef* provide an opportunity to undertake such analyses. The primary amino acid sequences of these proteins are known, and deletion-mapping studies can be performed, making it possible to define motifs that may be responsible for generating specific immune reactivity. In addition, known motifs promote targeting of proteins to specific cell compartments. These include myristoylation sites, signal sequences, and endoplasmic reticulum trafficking signals (glycosylation motifs). Through a combination of modeling, motif recognition, and the application of bioinformatics and empirical testing, it should be possible to better define the underlying principles that govern these responses. Ideally, such efforts may allow prediction of immune responses to specific amino acid sequences and improved efficacy in the design of effective immunogens for vaccine development. Such vaccines could be rapidly synthesized by recombinant DNA techniques and formulated to respond in a timely fashion to infectious outbreaks. Although this technology will not provide the first line of defense in containing viral infection, it may provide an opportunity to immunize populations more distant from sites of initial exposure and thus contain disease.

USE OF RECOMBINANT TECHNIQUES TO DEVELOP TREATMENTS

Finally, these recombinant techniques also provide the opportunity for more rapid development of antiviral treatment for human infectious disease. In the case of Ebola virus we have learned that expression of the viral glycoprotein can lead to severe cytopathic effects in cell culture models in this disease (Yang et al., 2000). Synthesis of the Ebola virus glycoprotein causes severe cellular cytotoxicity when expressed in endothelial cells in the laboratory. This glycoprotein mediates this effect through a specific domain of the protein. When this domain is deleted, its toxicity is eliminated. We have been able to map the genetic determinants that lead to cellular cytotoxicity rapidly, and it should be possible to use this information to identify the mechanism of such viral cytopathic effects. During infectious outbreaks where disease mortality is high, it may be possible to limit the mortality of the disease rather than eliminate the infection entirely to prevent its lethality in large populations. In the case of Ebola virus we have begun to identify drugs that interfere with the synthesis of this protein. Additional studies will provide further insight into its molecular mechanism. These mechanisms of cytotoxicity that apply to Ebola virus may well apply to other hemorrhagic fevers and could provide for generic classes of drugs that may be useful in blunting the lethality of an ongoing outbreak.

CONCLUSIONS

Molecular genetics, information technologies, and bioengineering can significantly improve efforts to prevent, detect, and treat infectious diseases. It is likely that improved technologies for data collection, molecular analysis, and recombinant genetics will facilitate our ability to contain outbreaks of both naturally occurring and iatrogenic infections. Implementation of cost-effective, rapid, high-throughput screening technology at remote sites with information transfer to diverse locations for data analysis by appropriate epidemiologic and scientific personnel can allow for more expedient and effective responses to emerging infectious threats.

REFERENCES

Bagarazzi, M. L., J. D. Boyer, V. Ayyavoo, and D. B. Weiner. 1998. Nucleic acid-based vaccines as an approach to immunization against human immunodeficiency virus type-1. Current Topics in Microbiology and Immunology, 226:107-143.

Baize, S., E. M. Leroy, M. C. Georges-Courbot, M. Capron, J. Lansoud-Soukate, P. Debre, S. P. Fisher-Hoch, J. B. McCormick, and A. J. Georges. 1999. Defective humoral responses and extensive intravascular apoptosis are associated with fatal outcome in Ebola virus-infected patients. Nature Medicine, 5:423-426.

Burton, D. R., and J. P. Moore. 1998. Why do we not have an HIV vaccine and how can we make one? Nature Medicine, 4:495-498.

Donnelly, J. J., J. B. Ulmer, J. W. Shiver, and M. A. Liu. 1997. DNA vaccines. Annual Review of Immunology, 15:617-648.

Gao, F., D. L. Robertson, S. G. Morrison, H. Hui, S. Craig, J. Decker, P. N. Fultz, M. Girard, G. M. Shaw, B. H. Hahn, and P. M. Sharp. 1996a. The heterosexual human immunodeficiency virus type 1 epidemic in Thailand is caused by an intersubtype (A/E) recombinant of African origin. Journal of Virology, 70:7013-7029.

Gao, F., S. G. Morrison, D. L. Robertson, C. L. Thornton, S. Craig, G. Karlsson, J. Sodroski, M. Morgado, B. Galvao-Castro, H. von Briesen, et al. 1996b. Molecular cloning and analysis of functional envelope genes from human immunodeficiency virus type 1 sequence subtypes A through G. The WHO and NIAID networks for HIV isolation and characterization. Journal of Virology, 70:651-667.

Heilman, C. A., and D. Baltimore. 1998. HIV vaccines—where are we going? Nature Medicine, 4:532-534.

Nabel, G. J. 1999. Surviving Ebola virus infection. Nature Medicine, 5:373-374.

Palella, F. J. 1998. Declining morbidity and mortality among patients with advanced HIV infection. New England Journal of Medicine, 328:853.

Sullivan, N.J., A. Sanchez, P.E. Rolling, Z. Yang, and G.J. Nabel. 2000. Development of a preventive vaccine for Ebola virus infection in primates. Nature, 408:605-609.

Tighe, H., M. Corr, M. Roman, and E. Raz. 1998. Gene vaccination: Plasmid DNA is more than just a blueprint. Immunology Today, 19:89-97.

UNAIDS, Joint United Nations Programme on HIV/AIDS. 1998. Report on the Global HIV/AIDS Epidemic. New York: UNAIDS.

Xu, L., A. Sanchez, Z. Yang, S. R. Zaki, E. G. Nabel, S. T. Nichol, and G. J. Nabel. 1998. Immunization for Ebola virus infection. Nature Medicine, 4:37-42.

Yang, Z., H. J. Duckers, N. J. Sullivan, A. Sanchez, E. G. Nabel, and G. J. Nabel. 1999. Identification of the Ebola virus glycoprotein as the main viral determinant of vascular cell cytotoxicity and injury. Nature Medicine, 6:886-889.

9

Next Steps in the Global Surveillance for Anti-Tuberculosis Drug Resistance

Ariel Pablos-Mendez

BACKGROUND

Antimicrobial resistance in previously susceptible organisms occurs wherever antibiotics are used for the treatment of infectious diseases in humans and animals (Neu, 1992). Because of increasing antibiotic use and misuse over the past decades, resistance has emerged in all kinds of microorganisms—including *Mycobacterium tuberculosis*—posing new challenges for both clinical management and control programs (Dooley et al., 1992; Kochi et al., 1993).

Resistance of *M. tuberculosis* to antibiotics results from artificial amplification of spontaneous mutations in the genes of the tubercle bacilli. Treatment with a single drug—due to irregular drug supplies, inappropriate prescription practices, or poor adherence to treatment—suppresses the growth of susceptible strains to that drug but permits the multiplication of drug-resistant strains. This phenomenon is called *acquired resistance*. Subsequent transmission of such resistant strains from an infectious case to other persons leads to disease that is drug resistant from the outset, a phenomenon known as *primary resistance*.

Dramatic outbreaks of multidrug-resistant tuberculosis in HIV-infected patients in the United States (Centers for Disease Control, 1991; Edlin et al., 1992) and Europe (Monno et al., 1991; Herrera et al., 1996) recently focused international attention on the emergence of strains of *M. tuberculosis* that are resistant to antimycobacterial drugs (Frieden et al., 1993). Multidrug-resistant tuberculosis—defined as resistance to the two most important antimicrobial drugs, isoniazid and rifampin—is a significant threat to tuberculosis control. Patients infected with strains resistant to multiple drugs are extremely difficult to cure, and the necessary treatment is much more toxic and expensive. Drug resistance is therefore a potential threat to the standard international method of tuberculosis control, the DOTS strategy, or "Directly Observed Treatment, Short-Course."

WORLD HEALTH ORGANIZATION/ INTERNATIONAL UNION AGAINST TUBERCULOSIS AND LUNG DISEASE GLOBAL PROJECT ON ANTI-TUBERCULOSIS DRUG RESISTANCE SURVEILLANCE

In early 1994 the Global Tuberculosis Program of the World Health Organization (WHO) joined forces with the International Union Against Tuberculosis and Lung Disease and started the Global Project on Anti-Tuberculosis Drug Resistance Surveillance. The objectives of the project were to measure the prevalence of anti-tuberculosis drug resistance in several countries worldwide using standard methods and to study the correlation between the level of drug resistance and treatment policies in those countries.

The first step toward achieving the objectives was the development in 1994 of common definitions and guidelines. These focused on three major principles: (1) surveillance must be based on a sample of tuberculosis patients representative of all cases in the country, (2) primary and acquired drug resistance must be clearly distinguished in order to interpret the data correctly, and (3) proper laboratory performance must be assured.

The second step entailed establishment of a global network of Supranational Reference Laboratories (SRLs) to serve as reference centers for quality assurance of drug susceptibility testing. The network comprised 22 SRLs in 1997 and is expanding. The third step involved organizing a working group under the leadership of the WHO with representatives of national tuberculosis programs and research institutions from more than 50 countries to implement surveillance projects at the national level.

The first phase of the global project gathered results from 35 countries on five continents. Surveillance activities were conducted on approximately 50,000 tuberculosis cases in areas representing 20 percent of the world's population. Each study enrolled on average 1,200 tuberculosis patients (the range was 59 to 14,344). Testing for the presence of isoniazid and rifampin was conducted, and resistance to ethambutol and streptomycin was evaluated. Overall, agreement between the results of the SRL and the various national reference laboratories (NRLs) was 96 percent. All but three countries were able to distinguish between primary and acquired resistance.

Main Findings, 1994 to 1997

Primary Drug Resistance

Information was obtained from cases with effectively no previous treatment, thus reflecting the transmission of strains that were already resistant. The prevalence of resistance to any drug ranged from 2 percent (Czech Republic) to 40 percent (Dominican Republic), with a median value of 9.9 percent. Primary resistance to all four drugs tested was found in a median of 0.2 percent of the cases (range of 0 to 4.6 percent). Primary multidrug-resistant tuberculosis was found in every country surveyed except Kenya, with a median prevalence of 1.4 percent and a range of 0 (Kenya) to 14.4 percent (Latvia).

Acquired Drug Resistance

The presence of acquired drug resistance reflects more recent case mismanagement. The populations assessed for this part of the study were patients who were treated for a month or longer in the past. As expected, the prevalence of acquired drug resistance was much higher than that of primary drug resistance. The prevalence of acquired resistance to any drug ranged from 5.3 percent (New Zealand) to 100 percent (Ivanovo Oblast, Russia), with a median value of 36 percent. Resistance to all four drugs among previously treated patients was reported in a median of 4.4 percent of the cases (range 0 to 17 percent). The median prevalence of acquired multidrug-resistant tuberculosis was 13 percent, with a range of 0 percent (Kenya) to 54 percent (Latvia).

Overview of the Global Situation

These findings are probably an underestimate of the magnitude of the problem worldwide because the countries surveyed had better than average tuberculosis control. Resistance to tuberculosis drugs is probably present everywhere in the world. Certainly, multidrug-resistant tuberculosis is present on all five continents; a third of the countries surveyed had levels of multidrug-resistant tuberculosis in more than 2 percent of new patients. In Latvia 30 percent of all patients presenting for treatment had multidrug-resistant tuberculosis. The region of Russia surveyed showed 5 percent of tuberculosis patients with multidrug-resistant tuberculosis. In the Dominican Republic 10 percent of tuberculosis patients had multidrug-resistant tuberculosis. In Africa the Ivory Coast has also witnessed the emergence of multidrug-resistant tuberculosis. Preliminary reports from Asia (India and China) show high levels of drug resistance as well. In the state of Delhi, India, 13 percent of all tuberculosis patients had multidrug-resistant tuberculosis.

Correlation of Prevalence of Drug Resistance with Tuberculosis Control Policy and Activities

An important finding of the study was the higher prevalence of multidrug-resistant tuberculosis in countries categorized by the WHO as having poor control programs. Similarly, the higher the proportion of retreatment cases (the result of a poor program), the higher the levels of drug resistance among new patients. The use of standardized short-course chemotherapy regimens, on the other hand, was associated with lower levels of drug resistance.

Conclusions of the First Phase of the Project

First, drug resistance is ubiquitous. The Global Project found it in all countries surveyed. The levels of resistance to isoniazid are high, and continued failure to improve tuberculosis control will fuel multidrug resistance. Second, there are several hot spots around the world where multidrug-resistant tuberculosis prevalence is high and could threaten control programs. These include Latvia, Estonia, and Russia in the former Soviet Union, the Dominican Republic and Argentina in the Americas; and the Ivory Coast in Africa. Preliminary reports from Asia also show high levels of multidrug-resistant tuberculosis. Urgent intervention is needed in these areas.

Third, there is a strong correlation between both the overall quality of tuberculosis control and the use of standardized short-course chemotherapy and low levels of drug resistance. A high prevalence of multidrug-resistant tuberculosis is the result of therapeutic anarchy. Half the countries or regions with the worst tuberculosis control had primary MDR levels above 2 percent, compared with one-fifth of those with moderate control and none of the countries with the highest standard of tuberculosis control.

Finally, the level of multidrug-resistant tuberculosis is a useful indicator of national tuberculosis program performance. As shown by the global project and by previous experiences in Korea and New York, the prevalence of primary multidrug-resistant tuberculosis is a good "summary" indicator of the performance of tuberculosis control programs in recent years.

LABORATORY ISSUES

The network of SRLs of the global project is a model of international scientific collaboration in support of an important public health initiative. The benefits brought to NRLs by this network will extend beyond the global project itself. However, the results of the project thus far merely update the pessimistic view expressed by Fox (1977) about drug susceptibility testing (1977). Twenty years later, as drug resistance is emerging as a clinical and an epidemiologic concern, several laboratories in the world are finally capable of providing accurate and reliable drug susceptibility testing results, especially for isoniazid and rifampin.

With the implementation of a quality assurance program, internationally comparable results of drug susceptibility testing could be obtained. When future surveys are planned in a given country, an SRL should first conduct proficiency testing of the responsible NRL and together develop a scheme for quality control (World Health Organization/International Union Against Tuberculosis and Lung Disease, 1997). When an NRL does not exist or its drug susceptibility testing performance remains suboptimal, it is preferable to have strains tested at the SRL itself. In addition to the accuracy of drug susceptibility testing, quality control needs to emphasize safety in the laboratory. For international transport of strains, adherence to international regulations is mandatory (World Health Organization, 1993).

One of the important findings of the global project is that all four laboratory methods for drug-susceptibility testing can achieve similar levels of accuracy (Canetti, 1965). Standard and economic variants of the proportion, the resistance ratio, the absolute concentration, and the radiometric BACTEC 460 methods have yielded similar drug susceptibility results, both between and within laboratories, for the four first-line anti-tuberculosis drugs evaluated. These findings justify the continued use of traditional drug susceptibility testing methods in institutions familiar with them or unable to afford more recent technology.

For efficient implementation of the global project, ideally, several regional networks should be established to conduct technical exchanges between the participating countries in the region. Geographical proximity and cultural similarity facilitate technical exchange and collaboration to implement local surveys. The regional offices of the WHO could play a critical role in the development of these networks by identifying target countries and experts in the region.

The Western Pacific Regional Office of the WHO established an organization of NRLs and SRLs in the Pacific region, modeled after the global SRL network. The Korean Institute of Tuberculosis carried out a quality assurance study on drug susceptibility testing in 1995 and 1996 in which the NRLs of China, Hong Kong, Malaysia, Thailand, and Vietnam participated. The results of the first round of proficiency testing, implemented by the Research Institute of Tuberculosis in Tokyo and the Korean Institute of Tuberculosis in Seoul, showed a fairly acceptable concordance, except for rifampin susceptibility testing. Three NRLs showed an unacceptable concordance rate (<80 percent) with RMP tests on the first round of testing but improved considerably (93 to 100 percent) in the second round. All of the participating NRLs showed an acceptable concordance rate for isoniazid susceptibility testing in both rounds of proficiency testing. Additional improvement in drug-susceptibility testing proficiency is expected as quality assurance at the regional level improves and expands.

Finally, although the global project almost exclusively used conventional epidemiologic and laboratory methodology, recent developments in molecular technology and their applicability in drug susceptibility testing and epidemiologic studies of multidrug-resistant tuberculosis must be acknowledged. Current tests can rapidly and directly ascertain the drug susceptibility status of a given strain in clinical specimens without culture (Telenti et al., 1993). We now know that the same genetic mutations underlie primary and acquired drug resistance, whether in AIDS patients or not, inside lung cavities or in extrapulmonary sites, or in Europe or sub-Saharan Africa (Cole, 1997).

An exciting approach to rapid drug susceptibility testing is the polymerase chain reaction (PCR)-based amplification of specific genes involved in resistance to individual drugs. These tests are approved for use only in smear-positive pulmonary tuberculosis, and the results must be examined in clinical context. The current limitations and expense associated with this technology prevent its use in routine clinical practice. An easier approach is to detect growth of *M. tuberculosis* in the presence of a given drug using mycobacteriophage (Jacobs et al., 1993). While these methods are promising for rifampin drug susceptibility testing, they have not been tested in field situations, and standardization is required before they can be used outside research laboratories.

Molecular techniques are also increasing our understanding of the epidemiology of tuberculosis and drug resistance (World Health Organization, 1995). Specific strains of *M. tuberculosis* can be identified by restriction fragment length polymorphism DNA fingerprinting (Thierry et al., 1990) and new DNA microarrays. In addition, clustering of strains sharing a DNA fingerprint suggests an increased probability of recent *M. tuberculosis* transmission (Daley et al., 1992). These molecular tools have provided refined descriptions of point-source tuberculosis outbreaks (Alland et al., 1994) as well as the dissemination of specific drug-resistant strains (Bifani et al., 1996). Genetic epidemiologists are also reexamining the century-old debate about recent transmission versus reactivation of latent infection in tuberculosis epidemiology (Friedman et al., 1995).

NEXT STEPS

The global project has highlighted the need for action in the areas of surveillance, management, and research.

Surveillance

As part of the global project, surveillance reports from 40 additional countries were made available in the year 2000, and trends over time will be available in some of the initial 35 participating countries. The following strategy is being followed:

- The well-established network of SRLs is a model for standardized surveillance of drug resistance and should be maintained as a global resource. Additional SRLs have been added, and regional networks are being consolidated.
- There will be adequate assessment of the level of multidrug-resistant tuberculosis in large countries (e.g., China, India, Russia), with expansion of surveillance activities beyond the regions studied. Areas not adequately covered during the first phase of the Global Project are being targeted.
- Future surveys are to collect and analyze individual data on age, HIV coinfection, and country of birth and on the contribution of unregulated private sectors to drug resistance.

Management

Countries without the DOTS tuberculosis control strategy need to implement it. This requirement is supported by the global project's finding of an association of low resistance and high-quality tuberculosis control. Previous experience has also demonstrated decreases in resistance, even in multidrug-resistant tuberculosis, following the introduction of DOTS tuberculosis control.

The global project does not directly address the issue of treatment regimens. Based on previous experience and the generally low levels of drug resistance, however, no alterations are yet required for the standard regimens recommended by the WHO and the International Union Against Tuberculosis and Lung Disease in new patients with tuberculosis (World Health Organization, 1996). For the management of multidrug-resistant tuberculosis, the reader is referred to "Guidelines for the Management of Drug-resistant Tuberculosis" (Crofton et al., 1997).

Research

The results of the global project have brought into focus several important questions. First, efforts must be made to assess the transmissibility and clinical virulence of multidrug-resistant tuberculosis compared to disease caused by drug susceptible strains. In vitro studies in the 1950s (Cohn et al., 1954) and recent epidemiologic evidence suggest that multidrug-resistant strains of *M. tuberculosis* may be less virulent or transmissible than drug-susceptible organisms. Modeling exercises could then put in perspective the impact of multidrug-resistant tuberculosis in global and regional trends.

Second, the global project has highlighted the need to define the impact of multidrug-resistant tuberculosis on treatment outcomes under program conditions in developing countries. In most settings with relatively low numbers of multidrug-resistant tuberculosis, current practice (standard regimens without drug-susceptibility testing) may still achieve target cure rates. However, in settings with an unusually high prevalence of multidrug-resistant tuberculosis, concerns have been raised about the risk of worsening the problem with the use standardized regimens vis-à-vis individualized regimens (based on drug-susceptibility testing). Beyond anecdotal evidence and small case series, evidence in support of that concern is lacking. Most treatment failures are due to reasons other than multidrug-resistant tuberculosis. While in individual patients resistance can only increase with additional waves of treatment, experience around the world shows that resistance levels in the community can be reduced by implementing sound control policies with standardized regimens (Kim and Hong, 1992; Chaulet, 1993). Theoretically, however, there must be a threshold of MDR levels above which standardized regimens will not work and that may indeed promote additional drug resistance. In 1999 the WHO formed a DOTS-Plus Working Group to address these issues.

Finally, pharmaceutical companies are urged to develop new antituberculosis drugs. Simple replacement of older products may not be a wise investment. The prime need for such drugs is to make DOTS strategies more efficient and to shorten the duration of treatment, thus making resistance less likely to emerge in the first place.

LABORATORY DRUG SUSCEPTIBILITY TESTING AND TUBERCULOSIS MANAGEMENT

Multidrug resistance may or may not become an insurmountable barrier to well-implemented tuberculosis control programs. On the other hand, patients with clinically active disease caused by multidrug-resistant tuberculosis face uncertain prospects of successful treatment (Goble et al., 1993; Mitchison and Nunn, 1986), side effects from medication (when available), and the associated expense of the medication (Mahmoudi and Iseman, 1993). Confronted with a potentially fatal disease, clinicians are ethically compelled to offer every available treatment. However, antibiotic misuse should not be justified by the (genuine) concerns of clinicians, a phenomenon that underlies the emergence of multidrug-resistant tuberculosis. In a program context, futile efforts made for a few incurable cases may only drain resources that could be used to cure many patients with drug susceptible tuberculosis and prevent multidrug-resistant tuberculosis in the first place.

Routine drug-susceptibility testing for clinical management of all tuberculosis patients is simply out of the question in most countries where the disease is concentrated. Even when available, such results do not affect the regimen used to treat over 95 percent of tuberculosis patients. Patients whose treatment fails after two courses of the standardized regimen are likely to harbor multidrug-resistant tuberculosis (50 percent or more; Kritski et al., 1997; Mazouni et al., 1992) and should be referred for expert management. A specialized unit may be regarded as an expensive luxury in many countries, but second-line drugs should be available only to such centers.

Some experts argue that in countries with a high tuberculosis burden and few resources, drug susceptibility testing should be done for surveillance purposes only, not to guide therapeutic decisions in individual patients. In such countries, drug susceptibility testing may distract from the essential duty to perform smear microscopy. Feedback of drug susceptibility testing results to individual physicians may be futile where second-line agents are not available or are unaffordable. The results of drug susceptibility testing in laboratories with a low volume of tests (<300 per year) may not even be accurate (Nitta et al., 1996). More importantly, drug susceptibility testing results may cause confusion and prompt inappropriate retailoring of therapeutic regimens without improving their efficacy (i.e., changing standardized regimens in patients with, for example, isoniazid monoresistance; Mahmoudi and Iseman, 1993). Poor implementation of recommended regimens may then lead to the therapeutic chaos in which multidrug-resistant tuberculosis thrives.

Policy makers and responsible clinicians should keep in mind that the cheapest multidrug-resistant tuberculosis treatment regimen is 100 times more expensive than the best first-line regimen (Crofton et al., 1997). Difficult choices must be made, since less than a third of patients with tuberculosis currently have access to adequate management. Countries that have secured these basic strategies may then decide to devote additional resources to fight multidrug-resistant tuberculosis. With qualified laboratories and available second-line drugs, drug susceptibility testing has been recommended for all new cases to help tailor the best possible therapeutic regimens under expert supervision (Centers for Disease Control and Prevention, 1993).

The WHO estimates the global incidence of tuberculosis at 8 million cases per year, with most occurring in India and sub-Saharan Africa (World Health Organization, 1999). Extrapolating results from the global project, 300,000 of these new cases are due to multidrug-resistant tuberculosis. New diagnostic technologies that simplify the tasks of surveillance programs and clinicians in less developed countries would be most welcome. Such methods should be acceptable in terms of cost, accuracy, speed, and complexity (direct based as opposed to culture based). In particular, new methods to detect microbial resistance to rifampin (practically equivalent to MDR) may revolutionize current recommendations for global surveillance and facilitate the development of therapeutic regimens appropriate to each country or region.

REFERENCES

Alland, D., G. E. Kalkut, A. R. Moss, R. A. McAdam, J. A. Hahn, W. Bosworth, E. Drucker, and B. R. Bloom. 1994. Transmission of tuberculosis in New York City: An analysis by DNA fingerprinting and conventional epidemiological methods. New England Journal of Medicine, 330:1710-1716.

Bifani, P. J., B. B. Plikaytis, V. Kapur, K. Stockbauer, X. Pan, M. L. Lutfey, S. L. Moghazeh, W. Eisner, T. M. Daniel, M. H. Kaplan, J. T. Crawford, J. M. Musser, and B. N. Kreiswirth. 1996. Origin and interstate spread of a New York City multidrug-resistant *Mycobacterium tuberculosis* clone family. Journal of the American Medical Association, 275:452-457.

Canetti, G. 1965. Present aspects of bacterial resistance in tuberculosis. American Review of Respiratory Disease, 92:687-703.

Centers for Disease Control and Prevention. 1991. Nosocomial transmission of multidrug-resistant tuberculosis among HIV-infected persons: Florida and New York, 1988-1991. Morbidity and Mortality Weekly Report, 40:585-591.

Centers for Disease Control and Prevention. 1993. Tuberculosis control laws—United States, 1993: Recommendations of the advisory council for the elimination of tuberculosis. Morbidity and Mortality Weekly Report, 42(RR-15):1-13.

Chaulet, P. 1993. Tuberculose et transition épidemiologique: le cas de l'Algerie. Annals of the Institut Pasteur, 4:181-187.

Cohn, M. L., C. Kovitz, U. Oda, and G. Middlebrook. 1954. Studies on isoniazid and tubercle bacilli: The growth requirements, catalase activities, and pathogenic properties of isoniazid-resistant mutants. American Review of Tuberculosis, 70:641-664.

Cole, S. T. 1997. Molecular basis of drug resistance in *M. tuberculosis*. The Wellcome Trust Meeting on Antibiotic Resistance, United Kingdom, May 18-21.

Crofton, J., P. Chaulet, D. Maher, et al. 1997. Guidelines for the Management of Drug-Resistant Tuberculosis. WHO/Gtuberculosis/96.210. Geneva: World Health Organization.

Daley, C. L., P. Small, G. Schecter, G. K. Schoolnik, R. A. McAdam, W. R. Jacobs, Jr., and P. C. Hopewell. 1992. An outbreak of tuberculosis with accelerated progression among persons infected with human immunodeficiency virus. New England Journal of Medicine, 326(4):231-235.

Dooley, S. W., W. R. Jarvis, W. J. Marlone, and D. E. Snider. 1992. Multidrug-resistant tuberculosis. Annals of Internal Medicine, 117:257-259.

Edlin, B. R., J. L. Tokars, M. H. Grieco, J. T. Crawford, J. Williams, E. M. Sordillo, K. R. Ong, J. O. Kilburn, S. W. Dooley, K. G. Castrio, et al. 1992. An outbreak of multidrug-resistant tuberculosis among hospitalized patients with the acquired immunodeficiency syndrome. New England Journal of Medicine, 326:1514-1521.

Frieden, T., T. Sterling, A. Pablos-Méndez, J. O. Kilburn, G. M. Cauthen, and S. W. Dooley. 1993. The emergence of drug-resistant tuberculosis in New York City. New England Journal of Medicine, 328:521-26.

Friedman, C., M. Y. Stoekle, B. N. Kreiswirth, W. D. Johnson, Jr., S. M. Manoach, J. Berger, K. Sathianathan, A. Hafner, and L. W. Riley. 1995. Transmission of multidrug-resistant tuberculosis in a large urban setting. American Journal of Respiratory and Critical Care Medicine, 152:355-359.

Fox, W. 1977. The modern management and therapy of pulmonary tuberculosis. Proceedings of the Royal Society of Medicine, 70:4.

Goble, M., M. D. Iseman, L. A. Madsen, D. Waite, L. Ackerson, and C. R. Horsburgh, Jr. 1993. Treatment of 171 patients with pulmonary tuberculosis resistant to isoniazid and rifampin. New England Journal of Medicine, 328:527-532.

Herrera, D., R. Cano, and P. M. K. Godoy. 1996. Multidrug-resistant tuberculosis outbreak on an HIV ward: Madrid, Spain, 1991-1995. CDC Morbidity and Mortality Weekly Report, 45:330-333.

Kim, S. J., and Y. P. Hong. 1992. Drug resistance of *Mycobacterium tuberculosis* in Korea. Tubercle and Lung Disease, 73:219-224.

Kochi, A., B. Vareldzis, and K. Styblo. 1993. Multidrug resistant tuberculosis and its control. Research in Microbiology, 144:104-110.

Kritski, A., L. S. Rodriguez de Jesus, M. K. Andrade, E. Werneck-Barroso, M. A. Vieira, A. Haffner, and L. W. Riley. 1997. Retreatment tuberculosis cases: Factors associated with drug resistance and adverse outcomes. Chest, 111:1162-1167.

Jacobs, W. R., R. G. Barletta, R. Udani, J. Chan, G. Kaikut, G. Sosne, T. Kieser, G. J. Sarkis, G. F. Hatfull, and B. R. Bloom. 1993. Rapid assessment of drug susceptibilities of *Mycobacterium tuberculosis* by means of luciferase reporter phages. Science, 260:819-822.

Mahmoudi, A., and M. D. Iseman. 1993. Pitfalls in the care of patients with tuberculosis. Journal of the American Medical Association, 270:65-68.

Mazouni, L., N. Zidouni, F. Boulahbal, and P. Chaulet. 1992. Treatment of Failure and Relapse Cases of Pulmonary Tuberculosis Within a National Programme Based on Short Course Chemotherapy: Preliminary Report. Tuberculosis Surveillance Research Unit of the International Union Against Tuberculosis and Lung Disease, Progress Report 1992 2:36-42.

Mitchison, D. A., and A. J. Nunn. 1986. Influence of initial drug resistance on the response to short-course chemotherapy of pulmonary tuberculosis. American Review of Respiratory Disease, 133:423-430.

Monno, L., G. Angarano, S. Carbonara, S. Coppola, D. Costa, M. Quarto, and G. Pastore. 1991. Emergence of drug-resistant *Mycobacterium tuberculosis* in HIV-infected patients. Lancet, 337:852.

Neu, H. C. 1992. The crisis of antibiotic resistance. Science, 257:1064-1073.

Nitta, A. T., P. T. Davidson, M. L. de Koning, and R. J. Kilman. 1996. Misdiagnosis of multidrug-resistant tuberculosis possibly due to laboratory-related errors. Journal of the American Medical Association, 276:1980-1983.

Telenti, A., P. Imboden, F. Marchesi, D. Lowrie, S. Cole, M. J. Colston, L. Matter, K. Schopfer, and T. Bodmer. 1993. Detection of rifampin-resistance mutations in *Mycobacterium tuberculosis*. Lancet, 341:647-650.

Thierry, D., M. D. Cave, K. D. Eisenach, J. T. Crawford, J. H. Bates, B. Gicquel, and J. L. Guesdon. 1990. IS*6110*, an IS-like element of *Mycobacterium tuberculosis* complex. Nucleic Acids Research, 18:188.

World Health Organization. 1993. Safe shipment of specimens and infectious materials. Pp. 48-54 in Laboratory Biosafety Manual, 2nd ed. Geneva: WHO.

World Health Organization. 1996. Treatment of Tuberculosis: Guidelines for National Programmes. WHO/tuberculosis/96.199. Geneva: WHO.

World Health Organization Global Tuberculosis Programme. 1995. The Use of Restriction Fragment Length Polymorphism (RFLP) Analysis for Epidemiological Studies of Tuberculosis in Developing Countries. Report of a WHO Review and Planning Meeting. WHO/tuberculosis/95.194. Geneva: WHO.

World Health Organization Global Tuberculosis Program. 1999. Global Tuberculosis Control: WHO Report. WHO/tuberculosis/99.250. Geneva: WHO *(www.who.int/gtb/publications /globrep99/index.html).*

World Health Organization/International Union Against Tuberculosis and Lung Disease. Global Working Group on Antituberculosis Drug Resistance Surveillance. 1997. Guidelines for Surveillance of Drug Resistance in Tuberculosis. WHO/tuberculosis/96.216. Geneva: WHO.

10

Antibiotic Discovery by Microarray-Based Gene Response Profiling

Gary K. Schoolnik and Michael A. Wilson

INTRODUCTION

The increasing availability of complete genome sequences of medically important bacteria has led to the need to extract functional information from the nucleotide sequence in a form that can be used to better understand pathogenesis and physiology and for the development of new diagnostics, antibiotics, and vaccines. Sequence annotation can initiate this process by identifying open reading frames (ORFs), followed by functional predictions of their putative protein products by reference to homologues of known function found in other organisms. This, however, is a nonempirical exercise, and most of the functional assignments must be considered as hypotheses that need to be experimentally tested.

Microarray-based expression profiling is the logical next step after completion of the annotation process and is currently the linchpin of functional genomics (Schena et al., 1998). It is a comprehensive method by which each ORF of the genome of an organism, printed as discrete ORF-specific spots on a surface, is interrogated simultaneously to identify genes selectively expressed by the organism under specific conditions of growth, in response to drugs and other inhibitors of metabolic and/or biosynthetic pathways, and during its growth in host tissues. As generally practiced, the method requires a surface representation of the organism's genome (the microarray) and labeled RNA or cDNA prepared from the organism. This system and the microarray experimental design are discussed below.

MICROARRAY FABRICATION AND HYBRIDIZATION

A microarray is a surface that contains representations of each ORF of a sequenced and annotated genome (Schena et al., 1995; 1996; 1998; DeRisi et al., 1998). The surface used, the method by which ORF-specific DNA is bound to the surface, and the overall arrangement of the array vary with the system employed. Of the several available formats, the most common, economical, and flexible—and the one used by the authors in their study of the expression response of *Mycobacterium tuberculosis* (MTB) to the antibiotic isoniazid (Wilson et al., 1999; Behr et al., 1999)—was developed by Patrick Brown and colleagues at Stanford University (Schena et al., 1995; 1996; Shalon et al., 1996; DeRisi et al., 1998).

This array format consists of a microscope slide whose surface contains an *x* by *y* matrix of printed spots, each spot containing a polymerase chain reaction (PCR)-derived amplicon that corresponds to all or part of an ORF of the sequenced genome. Thus, each ORF of the genome is represented on the array as a separate spot, its location designated by its *xy* address. Additional spots are added and correspond to internal controls that monitor the printing and hybridization steps. For example, the MTB array fabricated in the authors' laboratory contain spots corresponding to nearly all of the 3,924 ORFs predicted by the Sanger Center's MTB sequencing project, together with additional control spots (Behr et al., 1999; Wilson et al., 1999).

Arrays of this kind can be fabricated in an academic laboratory. The process is begun by identifying optimal forward and reverse primer pairs for each predicted ORF. For this purpose the authors have used the *Primer 3* software provided by the Whitehead Institute (*http://www-genome.wi.mit.edu/genome_software/other/primer3.html*), and for the MTB microarray the average length of the resulting amplicons is 520 bps. At Stanford University the identified primers are downloaded to the Protein and Nucleic Acid Core facility, and the primers are synthesized by a 96-well multiplex oligonucleotide synthesizer. Subsequently, the corresponding forward and reverse primers are combined and PCR amplified using a 96-well thermocycler. Each PCR product is analyzed by gel electrophoresis, and those primer pairs that fail to yield a single product of the expected size are either resynthesized or a new primer pair set is selected for the ORF in question. Once the full set of amplicons is validated and adjusted to a standardized concentration, each is printed onto a prepared microscope glass slide by a robotic printer, and the DNA is cross-linked to the surface.

The principal innovation in gene expression profiling that was introduced by Brown and colleagues is two-color hybridization (Schena et al., 1995). This method employs two populations of cDNAs that have been differentially labeled with two different fluorochromes—the cDNAs having been derived from RNA prepared from the same organism cultivated under or exposed to two contrasting conditions. The two differentially labeled populations of cDNAs are combined in equal masses, applied to the array surface, and allowed to hybridize to the corresponding ORF-specific targets. The array is then scanned, and the intensity of each label for each ORF-specific spot is quantified. These values are compared, yielding ratios that serve as a measure of the relative degree of expression or repression of each ORF for the two tested conditions.

In practice, for organisms growing in bulk culture, the culture is split, and one of the resulting cultures is experimentally manipulated to test the hypothesis in question while the other serves as the unaltered control. At prespecified time points, aliquots containing equal numbers of organisms are removed from the control, and experimental culture and "total" RNA are extracted from each. Each pool of total RNA is then separately reverse transcribed in a "first-strand" reaction using random oligonucleotide primers. The authors have found that it is not necessary to separate bacterial ribosomal RNA from mRNA prior to cDNA labeling and hybridization even though the former constitutes approximately 97 percent of the total RNA. During the reverse transcription reaction, a nucleotide labeled with one kind of fluorochrome is incorporated into one of the cDNA products, and a nucleotide labeled with a different fluorochrome is incorporated into the alternative cDNA product. The two differentially labeled cDNA populations are then combined for use in the hybridization reaction that is conducted on the array surface.

MICROARRAY EXPERIMENTAL DESIGN

The use of contrasting conditions as the principal experimental paradigm is designed to learn more about the genome-wide response of the organism in three general areas of interest: (1) physiological responses to changes in growth conditions (e.g., log-phase growth versus stationary-phase growth or changes in gene expression during a diauxic shift); (2) physicochemical parameters that simulate or could serve as signatures for a particular host environment (e.g., low pH/high pH, low O_2/high O_2, iron present/iron absent, or H_2O_2 present/H_2O_2 absent); and (3) as part of the drug discovery process (e.g., drug or metabolic inhibitor present/absent).

In general, of the thousands of ORFs monitored during experiments of this kind, the vast majority show no differential expression for any particular tested condition. A much smaller set of genes exhibit selective expression or repression, and of these many, but not all, can be plausibly associated with an adaptive response by the organism that is specific to the condition under study. This appears to be particularly true early in the time course and under conditions that do not induce a generalized stress response. However, no rationale can be adduced, based on current knowledge, for a fraction of the selectively induced or repressed genes. Some of these have no inferred function because, during the annotation process, they were not found to be homologous to genes of known function. For others, while their functions may be known or can be inferred, their regulation by the condition studied could not have been predicted by prior knowledge of the organism's biology. Unexpected results of this kind are perhaps the most interesting and provide the basis for new hypotheses. Accordingly, microarray expression profiling can be viewed as a hypothesis-generating exercise.

ANALYSIS OF MICROARRAY GENE RESPONSE PROFILE DATA

Bioinformatics, in its broadest sense, is required for microarray experimentation at three levels: (1) ORF identification and functional annotation of the nucleotide sequence, (2) image processing of the microarray scan, and (3) analysis of microarray data in order to identify genes that exhibit similar patterns of regulation and that may be functionally integrated as an adaptive response of the organism to the tested condition. The latter two are microarray specific.

Image processing is conducted using the crude signal intensities from each of the two fluorochrome-specific channels. Refinement of these signals comes from the use of local background measurements and determination of average signal intensity for each spot using custom-written software (Michael Eisen, Scanalyze, 1998, available at *http://rana.stanford.edu*).

Microarray data analysis software is very much an evolving field (Eisen et al., 1998; Tamayo et al., 1999; Toronen et al., 1999) and in the authors' opinion is the current rate-limiting step in the experimental process. Even for the relatively small genomes of prokaryotes, a typical microarray-based experiment generates large datasets. Consider, for example, the detection of MTB genes selectively expressed or repressed at four different time points upon exposure of the organism to three buffers that differ only with respect to three pH values: 5.8, 6.8, and 7.4—conditions likely to be encountered by the organism in different host compartments.

Each of the ~4,000 ORFs represented on the array will yield a numerical value representing the cognate gene's level of expression for each pH compared to each of the two other tested pH values. At a minimum, for each of the five time points (including $t = 0$), this experiment would generate two values for each of the three pairwise comparisons of pH; therefore, the number of gene expression data points would be: $4{,}000 \times 2 \times 3 \times 5 =$

120,000. After statistically derived threshold values are set that distinguish expression and repression from system noise and provide the metric for the amplitude of the observed changes, the basic question asked of such data is: Which genes are induced and which are repressed upon exposure of the organism to one condition compared to a second condition?

The biological presumption is that microarray-derived patterns of gene expression should correspond to regulatory networks of the organism that govern the adaptive response. For this purpose, clusters of genes that exhibit the same expression/repression behavior with respect to amplitude, vector, and time are identified. In turn, this information is further refined by reference to the putative or known functions of genes comprising discrete clusters including their relationship to metabolic or biosynthetic pathways that may be coordinately regulated by the condition under study. Membership in regulons of this kind can thus be inferred, and further analysis of their location and transcriptional orientation will likely show that some reside in operon-like gene clusters. Software currently available in the public domain for these and related applications includes the cluster analysis program of Eisen et al. (1998) and the self-organizing map-based program of Tamayo et al. (1999). The integration of gene cluster data for *Escherichia coli* with empirically derived information about the corresponding metabolic pathways is enabled by reference to the EcoCyc project (Karp et al., 1999, found at *http://ecocyc.PangeaSystems.com/ecocyc/ecocyc.html*).

MICROARRAY-BASED EXPRESSION PROFILING AND THE DRUG DISCOVERY PROCESS

New antibiotics are urgently needed to respond to the growing problem of bacterial strains resistant to one or more commonly used drugs and the imminent appearance of strains resistant to all available classes of drugs. To accelerate the drug discovery process, microarray expression profiling has been proposed as a method to identify new drug targets, to discover the target of an active compound whose mode of action is unknown, and as the basis for a high-throughput screen to identify active leads from large compound libraries.

The logic of this approach follows. The mode of action of most antibiotic classes is the inhibition of a vital metabolic or biosynthetic pathway. Inhibitors of this kind predictably cause a decrease in pathway products downstream of the point of inhibition and an accumulation of pathway precursors upstream of the site of inhibition. The resulting fluctuations in the abundance of products and precursors are sensed by the genome and result in increased expression of genes coding for pathway enzymes distal to the point of inhibition and decreased expression of genes proximal to the point of inhibition. Increased expression of genes in associated shunt pathways may be expected to occur as well, including those that degrade or expel toxic byproducts that have accumulated as a result of pathway inhibition.

Accordingly, exposure of an organism to a drug of unknown mode of action should elicit an expression profile that incriminates the affected pathway and perhaps even the target in the pathway (Wilson et al., 1999). Through the use of inhibitors that are selective for many such pathways it should be possible to identify signature profiles that are pathway specific and characteristic for an inhibitor's mode of action (Wilson et al., 1999). Such information could accelerate drug development because identification of the pathway and the target in the pathway could lead to the rational design, synthesis, and testing of a series of chemical derivatives of the original compound.

Alternatively, the use of pathway-specific inhibitors for gene-response profiling experiments will illuminate other components of the affected pathway beyond the target itself by causing changes in the expression of their cognate genes (Wilson et al., 1999). If pathway inhibition is lethal, inhibition of any of the critical pathway enzymes should be lethal as well. They, therefore, constitute new drug target candidates. This strategy is particularly attractive when mutations of the original target in a critical pathway have led to antibiotic resistance. In this case, expression profiling may lead to the identification of other targets in the same pathway.

To identify entries in large compound libraries that inhibit a specific preselected pathway, one or more of the genes coding for pathway enzymes in the pathway—identified through the gene-response profiling process described above—can be used to signal pathway inhibition during high-throughput screens. For this purpose the promoter of the selected pathway gene is fused to a gene that encodes a signal suitable for a high-throughput assay, such as luciferase, and the reporter construct is then introduced into a bacterium of the same species. The pathway-specific reporter strain is then dispensed into multiple microtiter wells. Each well then receives a different entry of the compound library, and the plate is scanned to identify wells that emit the signal, indicating that the pathway-specific promoter has been induced.

MICROARRAY GENE RESPONSE PROFILING TO IDENTIFY COMPOUNDS THAT BLOCK TISSUE-SPECIFIC ADAPTATIONS OF THE ORGANISM

It is widely recognized that the mode-of-growth of bacteria in nature, including in tissues during the infectious process, is vastly different than for the broth-grown organisms that are used for most antibiotic development purposes. In turn, this has led to the idea that an unexplored avenue for drug development is the identification of compounds that selectively inhibit pathways that are crucial for growth of the organism in tissues, in pus, or on the surfaces of implantable prosthetic devices. It is argued that drugs of this kind might not disrupt the normal flora of the host since at least some of the pathways required by an invading organism would not be required by the same organism growing as a commensal in a normally colonized host compartment. In contrast, most of the antibiotics in common use today dramatically alter the normal flora of the host during treatment.

Microarray gene response profiling offers an efficient way to identify tissue-specific bacterial genes, and the two-color hybridization system described above is particularly well suited to compare the gene expression profiles of the same organism as a commensal and an infecting agent. However, while the concepts are simple, the technological difficulty involved in obtaining sufficient RNA from organisms growing in tissues is far greater than from broth-grown organisms, in part because their numbers are likely to be relatively small and RNA from the host and other commensal bacteria may give rise to ambiguities and signal-to-noise problems. However, linear amplification methods and chemical methods to intensify the fluorescent signals of the labeled probes will likely overcome these problems in the near future. Additionally, knowledge of microbial metabolic pathways that are activated by organisms growing in tissues may lead to the identification of in vitro growth conditions that contain tissue-specific signals that will trigger responses that normally occur only at sites of infection. Tissue-simulating media of this kind could be used for high-throughput screens for the identification of compounds that act on organisms in tissues, although this screen would not necessarily exclude compounds that also act on commensal organisms.

STEPS IN THE MICROARRAY-BASED GENE RESPONSE PROFILING DRUG DISCOVERY PROCESS THAT COULD BE FACILITATED BY LABORATORY AUTOMATION

The pharmaceutical industry has developed innovative and rapid ways to construct libraries containing 10^5 to 10^6 structurally characterized compounds and robotics-based methods to screen these combinatorial libraries against indicator strains or enzymes. This experience and technical platform can be adapted for microarray expression profiling; the following functionally discrete steps are particularly amenable to laboratory automation:

A. *Function*: Identification of genes selectively induced or repressed by pathway inhibitors.

1. Genome sequencing and automation.
2. Microarray fabrication.
- Production of ORF-specific DNA targets.
- Printing of targets onto the surface, yielding the x by y array.
- mRNA preparation and production of labeled cDNA.
- Array hybridization.
- Scanning of the hybridized array.
- Image processing: deriving data for analysis.
- Image analysis to yield the gene response profile.
- Data mining: comparative analysis to extract biological meaning.

B. *Function*: Identification of entries in a combinatorial compound library that inhibit a preselected critical metabolic/biosynthetic target.

- Identify sentinel pathway genes.
- Prepare promoter-fusion constructs that will signal promoter activation, indicating pathway inhibition.
- Robotic delivery of reporter strains carrying the promoter-fusion constructs to microtiter wells.
- Robotic delivery of each entry in the compound library to microtiter wells containing the reporter strain.
- Microtiter plate incubation followed by signal detection.

SUMMARY

Microarray expression profiling will likely become a common component of the drug discovery process, where it may be used in an iterative manner in the stepwise progression from compound libraries to the identification of leads and the refinement of leads. It will not replace biochemical assays of drug structure and action; rather, its main effect will be to focus the screening process on vital pathways that mediate the adaptive metabolic, physiological, and pathogenic responses of the organism.

ACKNOWLEDGMENTS

The authors gratefully acknowledge Pat Brown and Joe DeRisi for their generous assistance and advice, members of the Schoolnik laboratory for their unstinting support, and the National Institutes of Health for providing funds for this project through grants AI35969 and AI44826.

REFERENCES

Behr, M. A., M. A. Wilson, W. P. Gill, H. Salamon, G. K. Schoolnik, S. Rane, and P. M. Small. 1999. Comparative genomics of BCG vaccines by whole genome DNA microarray. Science, 284:1520-1523.

DeRisi, J. L., V. R. Iyer, and P. O. Brown. 1998. Exploring the metabolic and genetic control of gene expression on a genome scale. Science, 278:680-686.

Eisen, M. B., P. T. Spellman, P. O. Brown, and D. Botstein. 1998. Cluster analysis and display of genome-wide expression patterns. Proceedings of the National Academy of Sciences USA, 95:14863-14868.

Karp, P. D., M. Riley, S. M. Paley, A. Pellegrini-Toole, and M. Krummenacker. 1999. Eco Cyc: Encyclopedia of *E. coli* genes and metabolism. Nucleic Acids Research, 27:55-58.

Schena, M., D. Shalon, R. W. Davis, and P. O. Brown. 1995. Quantitative monitoring of gene expression profiles with a complementary DNA microarray. Science, 270:467-470.

Schena, M., D. Shalon, R. Heller, A. Chai, P. O. Brown, and R. W. Davis. 1996. Parallel human genome analysis: Microarray-based expression monitoring of 1000 genes. Proceedings of the National Academy of Sciences USA, 93:1061-1069.

Schena, M., R. A. Heller, T. P. Theriault, K. Konrad, E. Lachenmeier, and R. W. Davis. 1998. Microarrays: Biotechnology's discovery platform for functional genomics. Trends in Biotechnology, 7:301-306.

Shalon, D., S. J. Smith, and P. O. Brown. 1996. A DNA microarray system for analyzing complex DNA samples using two-color fluorescent probe hybridization. Genome Research, 7:639-645.

Tamayo, P., D. Slonim, J. Mesirov, Q. Zhu, S. Kitareewan, E. Dmitrovsky, E. S. Lander, and T. R. Golub. 1999. Interpreting patterns of gene expression with self-organizing maps: Methods and applications to hematopoietic differentiation. Proceedings of the National Academy of Sciences USA, 96:2907-2912.

Toronen, P., M. Kolehmainen, G. Wong, and E. Castren. 1999. Analysis of gene expression data using self-organizing maps. FEBS Letters, 451:142-146.

Wilson, M. A., J. L. DeRisi, H. H. Kristensen, P. Imboden, S. Rane, P. O. Brown, and G. K. Schoolnik. 1999. Exploring drug-induced alterations in gene expression in *Mycobacterium tuberculosis* by microarray hybridization. Proceedings of the National Academy of Sciences USA, 96(22):12833-12838.

11

Sequencing Influenza A from the 1918 Pandemic, Investigating Its Virulence, and Averting Future Outbreaks

Jeffery K. Taubenberger

INTRODUCTION

In 1918 and 1919, an influenza pandemic of unprecedented virulence swept the globe, leaving up to 40 million people dead in its wake (Crosby, 1989). Although the virus responsible for this catastrophe was not isolated at the time, it is now possible to study the genetic features of the 1918 virus using fixed and frozen tissue specimens recovered from victims of the pandemic (Taubenberger et al., 1997; Reid et al., 1999). The study of the 1918 virus is not just one of historical curiosity. Because influenza viruses continually evolve by mechanisms of antigenic shift and drift, new influenza strains—as emerging pathogens—continue to threaten human populations. Pandemic influenza viruses have emerged twice since 1918, in 1957 and 1968, and the risk for future influenza pandemics is thought to be high. An understanding of the genetic makeup of the most virulent influenza strain in history may facilitate the prediction and prevention of future pandemics.

HISTORY OF THE 1918 PANDEMIC

The influenza pandemic of 1918 was exceptional in both its breadth and its depth. The first wave of influenza in the spring and summer of that year was highly contagious but caused few deaths. But in late August a virulent form of the disease emerged and swept the globe in 6 months. The main wave of the global pandemic occurred in September through November of 1918, killing more than 10,000 people each week in some

U.S. cities. Outbreaks of the disease spread across not only North America and Europe but also as far as the Alaskan wilderness and the most remote islands of the Pacific. Large proportions of the population became ill, with 28 percent of the U.S. population estimated to have been infected. The disease was also exceptionally severe, with mortality rates greater than 2.5 percent of the U.S. population, compared to less than 0.1 percent in other influenza epidemics. Incredibly, some isolated populations had mortality rates of over 70 percent (Crosby, 1989).

Furthermore, in the 1918 pandemic most deaths occurred among young adults, a group that usually has a very low death rate from influenza. Influenza and pneumonia death rates for 15 to 34 year olds were more than 20 times higher in 1918 than in previous years, with 99 percent of excess deaths occurring among people under age 65. It has been estimated that the 1918 influenza epidemic killed 675,000 Americans, including 43,000 servicemen mobilized for World War I. The impact was so profound that it depressed the average life expectancy in the United States by more than 10 years (Reid and Taubenberger, 1999; Linder and Grove, 1947).

INFLUENZA A VIRUSES

Influenza A viruses are negative-strand RNA viruses with a segmented genome with eight gene segments coding for 10 proteins. They are known to infect a wide variety of warm-blooded animals, including birds and mammals. Antibodies against the hemagglutinin (HA) protein prevent receptor binding and are very effective at preventing reinfection with the same strain. HA can change in order to evade previously acquired immunity either by antigenic *drift*, whereby mutations of the currently circulating HA gene disrupt antibody binding, or by antigenic *shift*, in which the virus acquires an HA of a new subtype. Fifteen HA subtypes of influenza A are known to exist in wild birds (H1 through H15) and provide a source of HAs that are novel to humans (Webster, 1997). Both the 1957 and 1968 pandemics resulted from a shift in HA, and in both cases the hemagglutinins of the pandemic strains were closely related to avian strains (Bean et al., 1992; Schafer et al., 1993). For a pandemic influenza strain to emerge, the virus must have a hemagglutinin antigenically distinct from the one currently prevailing; this hemagglutinin cannot have circulated in humans for the past 60 to 70 years; and the virus must be transmissible from human to human (Kilbourne, 1997). The other surface protein, neuraminidase (NA), which has nine described subtypes (N1 through N9), also has been shown to shift in pandemics.

The natural reservoir for influenza virus is thought to be wild waterfowl. Periodically, genetic material from avian strains emerges in strains infectious to humans, and because pigs can be infected with both avian and human strains, they are thought to be an intermediary in this process. In 1979 an avian influenza A virus (without reassortment) entered the swine population in northern Europe, forming a stable viral lineage (Scholtissek et al., 1983). Influenza strains with recently acquired genetic material were responsible for pandemic influenza outbreaks in 1957 and 1968. Until recently there was no evidence that a wholly avian influenza virus could directly infect humans. However, in Hong Kong 18 people were infected with an avian H5N1 influenza virus in 1997, and six died of complications after infection (Subbarao et al., 1998; Claas et al., 1998). To understand the emergence of pandemic influenza strains, it is extremely important to determine how influenza viruses move between species. Learning more about the relationship of the 1918 influenza virus to swine and avian viral strains is one of the primary goals of this project.

SEQUENCING INFLUENZA A FROM THE 1918 PANDEMIC

The broad goal of research in this area has been twofold: (1) Where did the 1918 influenza virus come from, and how did it infect people? (2) Are there any genetic features of the sequence that would provide insight into the virulence of this strain? In total, 78 autopsy cases of victims of the lethal fall wave of the 1918 pandemic were examined for this study. Seventy-four of these consisted of fixed tissues. The majority of these individuals died of secondary bacterial pneumonia. Because they often had clinical courses longer than 1 week, it was extremely unlikely that any of the tissue samples from these cases would still retain influenza RNA. However, a subset of individuals died within 1 week with very unusual and characteristic lung pathology, including massive pulmonary edema or hemorrhage. While these pathological changes have occasionally been observed in other influenza outbreaks (including the 1957 flu), their prominence in 1918 is one of the cardinal features of the "Spanish" flu. It was this subset of patients on which research efforts were concentrated. Three influenza RNA positive cases were identified (Taubenberger et al., 1997; Reid et al., 1999).

While the serological data suggested that the 1918 hemagglutinin would resemble the H1-subtype swine flu isolated in 1930, the avian origin of the 1957 and 1968 HAs made it possible that the 1918 H1 would more closely resemble an avian H1. The complete coding sequence of the gene was generated from the South Carolina case and was confirmed using RNA from the New York and Alaska cases. Out of the 981 bases of the HA1 domain of the gene, only two nucleotide differences were noted between the cases. One of these differences (in the New York case) would change an amino acid as compared to the sequence of the South Carolina and Alaska cases. Interestingly, that change occurs at one of the critical amino acids involved in receptor binding. The overall receptor binding pattern for the 1918 hemagglutinin is most similar to those of classic swine influenza strains, and it is possible that the New York case could specifically bind both avian- and mammalian-type receptors, a property it would share with classic swine influenza viruses (Reid et al., 1999).

The full-length sequence of the 1918 HA shows that it is most closely related to the human and swine influenza strains of the 1930s. While it is more closely related to avian strains than any subsequent mammalian H1, it is phylogenetically distinct from current avian H1s (Reid et al., 1999). It is probable that the HA involved in the pandemic did not pass directly from an avian source to its pandemic form but rather spent some unknown amount of time adapting in a mammalian host, although whether that host was human or swine is unclear. Phylogenetic analyses of the full-length 1918 HA sequence consistently place it in the mammalian clade, sometimes near the root of swine strains and sometimes near the root of human strains. Differences in placement probably reflect differences in mutation rates between swine and human strains. The most likely interpretation of these results is that the 1918 strains are most closely related to the common ancestor of all subsequent human and swine H1 strains.

An alternative hypothesis is that the 1918 virus may have acquired its HA by shift immediately before the pandemic from an avian virus. If avian influenza genes are in evolutionary stasis, as has been suggested (Murphy and Webster, 1996), such a virus would not resemble current avian strains phylogenetically. For the 1918 strain to be the result of the direct introduction of an avian hemagglutinin, antigenic drift would have to have occurred within the avian clade over the last 80 years. However, because there are no known samples of avian H1 viruses from 1918, this hypothesis cannot be tested.

The different mortality patterns among influenza pandemics indicate that influenza viruses might emerge in different ways. The pandemic of 1890 demonstrated a mortality pattern different from that of 1957 or 1968. While morbidity was high in each year from 1890 to 1892, mortality was low in 1890, rose in 1891, and peaked in 1892 (Glezen, 1996). In 1957 and 1968 mortality was highest in the first year of the pandemic. In light of the 1890 pattern it is intriguing that prior to the 1918 pandemic the mortality rate from influenza and pneumonia began to rise in 1915 and 1916. The rate dipped slightly in 1917 and then rose sharply with a classic "herald" wave in the spring of 1918; it finally skyrocketed with the most virulent form in the fall and winter of 1918 to 1919. Is it possible that a poorly adapted H1N1 was beginning to spread in 1915 that caused some serious illness but that was not yet perfectly adapted? The best argument against this scenario is that, if a strain with a new hemagglutinin was causing enough illness to affect the national death rates from pneumonia and influenza, it should have caused a pandemic sooner, and significant numbers of people should have been immune, or at least partially immunoprotected, in 1918. Eighty years later it is impossible to distinguish fluctuations in mortality caused by drift in the previous strain (H2N2, H3N8?; Webster et al., 1992) from early waves of a newly emerging virus. Both the 1957 and the 1968 pandemics were preceded by mild waves early in the same year, and there is evidence that the 1968 pandemic virus had begun to circulate several years earlier (Monto and Maassab, 1981). The reassorted viruses of 1957 and 1968 had human-adapted internal proteins; perhaps after a surface protein shift neither virus required a long adaptation period before causing a pandemic.

A number of scenarios for the origin of the 1918 flu are possible. First, it could have been an entirely avian virus that entered the human population in 1918 already capable not only of infecting people—as was the 1997 Hong Kong "chicken flu"—but also of spreading from human to human with extreme efficiency. Second, a wholly avian virus could have entered the human population some years before 1918, gradually establishing itself and adapting toward efficient replication and transmission in humans. Third, it could have been a reassortant virus with some genes of avian origin and some of human origin, as were the pandemics of 1957 and 1968. Fourth, the pandemic may have resulted from mutations in a previously circulating human virus that made it completely unrecognizable antigenically. The extent of the pandemic and the fact that it arose in a mild wave followed by more severe waves make the fourth possibility the least likely. The supreme efficiency with which the 1918 influenza spread and the evidence that many different genes contribute to host specificity argue against an immediate avian origin. Distinguishing between a wholly avian virus that had been adapting in humans for some years and a reassortant virus where some genes were of immediate avian origin and some from the previously circulating human strain will be difficult, given the lack of any contemporary avian or pre-1918 human strains for comparison. Reassortment has been the mechanism for two well-characterized pandemics, while a gradually adapting avian virus has never been detected.

Little is known about how genetic features of influenza viruses affect virulence. Virulence of a particular influenza strain is complex, involving several features that include host adaptation, transmissibility, tissue tropism, and replication efficiency. The genetic basis for each of these features is not yet fully characterized but is most likely polygenic in nature. There are, however, several identified mutations that do radically change the behavior of a given flu strain. In the case of the 1918 virus, neither the HA cleavage site mutation (Taubenberger et al., 1997; Reid et al., 1999) nor the Δ146 mutation of WSN/33 was present (Goto and Kawaoka, 1998; Taubenberger, 1998). Both of these changes have been shown to affect tissue tropism of the viral strain.

Fragmentary sequences of all the remaining gene segments of the 1918 virus have already been deciphered, and full-length segment sequences will be completed. Such sequences will allow complete phylogenetic analyses of each segment and will help elucidate the origin of the 1918 virus. Whether any particular genetic features of the virus can be related directly to its exceptional virulence is yet unclear. Even as the genetic structure of the Spanish flu virus is becoming fully known, other questions, such as the role that differences in immunity in different age groups played in the 1918 mortality, may be forever lost to study. It is hoped that knowledge gained by studying this very successful human pathogen can be applied to prevent or at least predict the emergence of new influenza viruses that have pandemic potential.

WAS THE 1918 FLU PANDEMIC A ONE-TIME EVENT OR ARE SIMILAR PANDEMICS FORTHCOMING?

The answer to this question is unclear. Because whatever viral, human, and historical factors existed in 1918 that allowed the emergence of such a pandemic, it remains possible, given a similar set of circumstances, that a pandemic of similar magnitude could reoccur. Additionally, the rapid movements of people around the globe make containment of such a newly emerging flu virus inconceivable.

Three factors were associated with virulence of the 1918 flu: viral, historical, and human. In terms of viral factors, the strain probably reflected an HA shift to H1 before or during 1918. The 1890 pandemic is thought by archeserological analysis to have been an H2N2 strain. Similarly, it has been proposed that an H3 subtype circulated in humans in 1900 (Webster et al., 1992). There is no evidence of pantropicity by viral mechanism and none is reflected in the pathology. Sequence analysis has revealed that the 1918 strain lacked the HA cleavage site mutation and the Δ146 mutation in NA. Little is understood about the genetic basis of virulence in influenza. Therefore, questions such as what virulence factors are in other genes and which genes shifted to form the pandemic virus will be difficult to address unless pre-1918 viral samples are recovered.

In terms of historical factors, did enhanced virulence occur solely because of World War I? This is doubtful because morbidity and mortality rates for both civilian and military populations were similar across the globe. However, the density of troops in camps and the increased movement of people because of the war did facilitate spread of the virus. The lack of effective vaccines, antibiotics, and antiviral drugs as well as inadequate nursing and hospital care certainly played a role in the morbidity and mortality of the pandemic.

Finally, purely viral factors are unlikely to account for markedly different age mortality rates. Therefore, human factors, both immunological and physiological, may have played a role in virulence: What was the pre-1918 exposure (by age group) to H1N1 influenza strains and other subtypes of influenza A? What other physiological age differences exist? Possibilities include changes in ligand specificity and distribution in the respiratory tree, changes in protease levels, and age-related response to primary influenza infection.

ROLE OF BATCH SCIENCE IN PREDICTING OR AVERTING A NEW INFLUENZA PANDEMIC

A three-pronged strategy is proposed to meet the challenge of predicting or averting a future influenza pandemic: (1) studying the basis of virulence of the 1918 flu directly by genetic and functional analyses, (2) studying the basic biology of flu and host responses in other flu strains to identify features associated with virulence, and (3) increasing surveillance of influenza strains in humans and domestic and wild animals.

Much larger inventories of data are needed for an increased understanding of influenza ecology. Because birds are thought to be a natural reservoir for influenza viruses, the following questions need to be addressed more thoroughly: What is the subtype distribution among species and locations (both domestic and feral)? What is the drift rate of avian viruses? What is the incidence of genetic shift between/among avian strains? Longitudinal sampling studies are needed to answer these questions.

Because influenza is a zoonotic disease, massively increased surveillance of human exposure to animal-adapted flu strains would help elucidate the mechanism for the emergence of novel (potentially pandemic) strains in humans. Were the recent Hong Kong H5N1 and H9N2 outbreaks and human/swine H3N2 reassortant outbreak in United States swine unique and unusual events or were they "surveillance artifacts" resulting from limited sampling and limited reagents? What is the incidence of human exposure to avian, swine, or equine viruses (e.g., in poultry workers and swine and horse farmers)? What are the clinical and subclinical infection rates? What is the incidence of interpandemic shifts among circulating human strains (e.g., shift of internal genes)? Current testing procedures using reagents developed for known circulating flu strains are not adequate to address these questions. Again, longitudinal sampling studies are needed.

The opportunity exists to collect and analyze large amounts of data currently unavailable for analysis to help elucidate the mechanisms by which influenza viruses can adapt and cause pandemic infections in humans. This challenge is clearly worth attempting, given the legacy of the 1918 Spanish flu.

REFERENCES

Bean, W., M. Schell, J. Katz, Y. Kawaoka, C. Naeve, O. Gorman, and R. Webster. 1992. Evolution of the H3 influenza virus hemagglutinin from human and nonhuman hosts. Journal of Virology, 66:1129-1138.

Claas, E. C., A. D. Osterhaus, R. van Beek, J. C. De Jong, G. F. Rimmelzwaan, D. A. Senne, S. Krauss, K. F. Shortridge, and R. G. Webster. 1998. Human influenza A H5N1 virus related to a highly pathogenic avian influenza virus. Lancet, 351:472-477.

Crosby, A. W. 1989. America's Forgotten Pandemic: The Influenza of 1918. Cambridge: Cambridge University Press.

Glezen, W. 1996. Emerging infections: Pandemic influenza. Epidemiologic Reviews, 18:64-76.

Goto, H., and Y. Kawaoka. 1998. A novel mechanism for the acquisition of virulence by a human influenza A virus. Proceedings of the National Academy of Sciences USA, 95:10224-10228.

Kilbourne, E. 1977. Influenza pandemics in perspective. Journal of the American Medical Association, 237:1225-1228.

Linder, F. E., and R. D. Grove. 1947. Vital Statistics Rates in the United States: 1900-1940. Washington, D.C.: Government Printing Office, p. 254.

Monto, A. S., and H. F. Maassab. 1981. Serologic responses to nonprevalent influenza A viruses during intercyclic period. American Journal of Epidemiology, 113:236-244.

Murphy, B. R., and R. G. Webster. 1996. Orthomyxoviruses in Virology, (3rd ed.), B. N. Fields, ed. Philadelphia: Lippincott-Raven, p. 1419.

Reid, A. H., T. G. Fanning, J. V. Hultin, and J. K. Taubenberger. 1999. Origin and evolution of the 1918 "Spanish" influenza virus hemagglutinin. Proceedings of the National Academy of Sciences USA, 96:1651-1656.

Reid, A. H., and J. K. Taubenberger. 1999. The 1918 flu and other influenza pandemics: "Over there" and back again. Laboratory Investigation, 79:95-101.

Schafer, J. R., Y. Kawaoka, W. J. Bean, J. Suss, D. Senne, and R. G. Webster. 1993. Origin of the pandemic 1957 H2 influenza A virus and the persistence of its possible progenitors in the avian reservoir. Virology, 194:781-788.

Scholtissek, C., H. Burger, O. Kistner, and K. F. Shortridge. 1983. Genetic relatedness of hemagglutinins of the H1 subtype of influenza A viruses isolated from swine and birds. Virology, 129:521-523.

Subbarao, K., A. Klimov, J. Katz, H. Regnery, W. Lim, H. Hall, M. Perdue, D. Swayne, C. Bender, J. Huang, M. Hemphill, T. Rowe, M. Shaw, X. Xu, K. Fukuda, and N. Cox. 1998. Characterization of an avian influenza A (H5N1) virus isolated from a child with a fatal respiratory illness. Science, 279:393-396.

Taubenberger, J. K., A. H. Reid, A. E. Krafft, K. E. Bijwaard, and T. G. Fanning. 1997. Initial genetic characterization of the 1918 "Spanish" influenza virus. Science, 275:1793-1796.

Taubenberger, J. K. 1998. Cleavage of influenza virus hemagglutinin into HA1, HA2: No laughing matter. Proceedings of the National Academy of Sciences USA, 95:9713-9715.

Webster, R. G., W. J. Bean, O. T. Gorman, T. M. Chambers, and Y. Kawaoka. 1992. Evolution and ecology of influenza A viruses. Microbiological Reviews, 56:152-179.

Webster, R. G. 1997. Predictions for future human influenza pandemics. Journal of Infectious Disease, 176(Suppl. 1):S14-S19.

PART II

FOOD SUPPLY

12

Ensuring Safe Food: An Organizational Perspective

John C. Bailar III

INTRODUCTION

The food supply in the United States is generally secure and relatively safe from noxious agents. It is also ample, inexpensive, varied, nutritious, and often appetizing—attributes that must be protected and extended at the same time that food safety is enhanced. Although poor nutrition is common, it is usually the result of failures at the individual and family levels rather than general institutional failures.

Furthermore, the United States, like few other nations, is fortunate to have no history of widespread famine, whether natural or man-made, or pestilential disease, at least for many decades. Thus, it may be difficult to imagine major disruptions in food production, transportation, or distribution. Simultaneous failures of major crops across this vast and varied country seem highly unlikely.

Yet we do have some food safety problems. Recent examples include parasites on imported raspberries, large-scale tainting of hamburger and milk, pathological organisms in cheese, and the recognition of uncommon but highly dangerous strains of *Escherichia coli.* Fortunately, most of these outbreaks were detected before large numbers of people were severely harmed.

THE MEANING OF SAFETY

What does it mean to say that our food supply is safe? Because perfect safety in our food supply is simply unattainable, an operational definition of safety is needed. Lowrance (1976) once proposed that "a thing is safe if its risks are deemed acceptable." This view makes safety a subjective matter, and it also makes clear that safety is a matter of degree, involving possible tradeoffs with other, perhaps beneficial, effects. Using Lowrance's definition, I would conclude that the U.S. food supply is safe. Nonetheless, it could be made safer.

CHANGES IN THE FOOD SUPPLY

The U.S. food supply changed enormously between 1900 and 1950. During this period a shift occurred away from subsistence farming; the use of preservatives (including cold and freezing) increased; and consumption of meat, eggs, and cheese rose. However, the food supply changed more between 1950 and 1999 than it did in the first half of the century. More recent changes include the growth in agribusiness and the consolidation of food processors, the loss of biodiversity (variety in the biological strains of plants and animals used for food), an explosive increase in meals eaten away from home (or at least prepared elsewhere), rising imports of many different fruits and vegetables, the changing mix of nutritional components—including an increase in protein and a decrease in fat consumption—and decreased needs for food energy per capita as fewer people engage in heavy manual labor. Many of the items we expect to see on our grocery shelves today were not available in 1950.

Changes in the food supply and our demands on it will continue and may even accelerate in directions that cannot be predicted. The food supply in 2050 likely will be quite different from that of today, and such change will require the recognition of new food-related problems as well as the development of new solutions, all on a large scale. In particular, a pressing need will emerge for changes in food inspection and in the assurance of food safety.

Many factors could have important effects on the U.S. food supply over the next 50 years. One is the resurgence of plant and animal pathogens, including new strains and new diseases that affect farm animals or food plants. Monoculture (the exclusive cultivation of a single, common, uniform product across a substantial area) has already replaced the family farm, which once produced several different kinds of food. Because of this specialization, there is cause for increased concern about food acting as a vector that can carry toxins or pathogens into the human gastrointestinal tract.

Population growth, which will create increasing demands for sheer quantity in the food supply, must also be considered, as should the depletion of resources such as fertilizers and energy that are needed to sustain agricultural productivity. The total area and the condition of agricultural lands also are important concerns. Agricultural land is diminishing, partly as a result of urban growth and sprawl; partly as a result of erosion, contamination with toxic chemicals, and other human-related activities; and partly as a result of the depletion of important nutrients in soil that has been tilled improperly or for too long. Water supplies are already tight and will become tighter as the available water must be shared by those who need it for agriculture and industry and by a growing urban and suburban population. Although new sources of fresh water may be developed—including economic approaches to desalinization—at the moment the expense of producing fresh water is likely to add substantially to the cost of agricultural products. In addition, major changes in climate may occur that could affect food production. Global climate change or environmental degradation also could have severe and essentially permanent effects.

Political unrest is unpredictable and could affect food sources in the United States as well as in countries that supply our imports. A point of special interest is the malicious release of harmful agents—whether microbiological, chemical, or physical—into the food supply. While there is no sign that this is currently a threat, the danger is real.

Several favorable influences may counterbalance these difficulties, including higher productivity from new strains of plants already in use and the genetic engineering of new plant species to produce food. We should expect to see new food products and new processes in ever-accelerating profusion. In its original form the green revolution is almost complete, but other research-based improvements in productivity may have profound effects on the assurance of a diverse and ample food supply.

The effects of some other changes in the food supply are more difficult to predict. Improvements in our understanding of nutrition may lead to demands for new kinds of food that will require novel approaches to agriculture and new types of food processing. Improvements in standards of living are generally positive, but they can and have led to harmful dietary patterns. Finally, it seems likely that our sense of what is safe (i.e., of what risks are acceptable) may change substantially. Such change is likely to be in the direction of diminished tolerance of problems and the resulting tightening of surveillance and control of the food stream. In short, the opportunity for "big" problems in our food supply remains very much with us, and the risks may have even increased in recent years because of the following trends:

- Industrialization of food production, in which an error can become a very large problem.
- The growth of feed lots for cattle, the presence of mega-swine

operations, and the centralization of food processing (although the closing of big-city stockyards is a counter example).

- Globalization of the food supply, including imports that are virtually uninspected.
- Possible terrorism (which might be far more effective than in an economy based on subsistence farming).
- Rapid, frequent global commerce and travel, which can lead to serious and widespread problems involving food-related organisms.

Three additional social problems are likely to exacerbate the effects of these trends:

- Denial—there seems to be a widespread belief that "it can't happen here."
- Inertia—including turf battles within or among regulatory agencies at all levels.
- Scientism—that is, the arrogance of those who think they have the knowledge and expertise to deal with any problems that may arise.

In this context, I think of Edmund Burke, who was reported to have said: "Good public policy is what men of good will, ten years hence, will wish we had done" (Stanlis, 1968). What will our children, grandchildren, and great-grandchildren wish we had done regarding future food supplies? There are many unknowns about the future of the food supply, and we must identify and prepare as best we can for those unknowns—whether good or bad and whether soon or well into the future.

FOOD INSPECTION AND TESTING

How much benefit has the present system of food inspection brought us? Have numerous threats that would otherwise be broad and common been kept in check or even eliminated? This question cannot be answered at this time, but we do know that traditional organoleptic inspection (inspecting by appearance, feel, and smell) does little to detect the chemical and biological threats of current concern, although it does offer other advantages, and that major chemical and biological threats could escape detection indefinitely.

The traditional inspection of animal and poultry carcasses may once have served the important purposes of public health, but current threats are posed by microbiological agents and chemical hazards that cannot be detected by the traditional means of visual inspection, touch, and smell. To be sure, meat and poultry inspectors have numerous other responsibilities (such as assuring food cleanliness, proper labeling, and humane slaughter), but they can add little to the prevention of modern foodborne health problems.

Our present testing program for chemicals in foods is grossly inadequate for understanding possible new threats or even for monitoring the food supply for existing threats. For example, in 1998 the Food Safety and Inspection Service Chemical Residue Program was designed to perform a total of 28,970 chemical tests on meat, but these samples were, by design, taken from a much smaller number of animals. The meat supply was divided into "slaughter classes" (such as broiler chickens) with an average of about 300 carcasses examined per class. The largest number of samples from any class was 460. These sample sizes not only are small but also are gathered from across the country over the course of an entire year. A total of 460 chickens, among the 7.5 billion slaughtered each year, represents about 1 in 16 million chickens sampled. This sample is not optimal for several other reasons: (1) there is no place-to-place or year-to-year linkage of findings, (2) sampling rates for establishments are proportional to their size (which has some important advantages in cost reduction but means that smaller places may escape monitoring), (3) the first stage of cluster sampling is random, and the second stage is a convenience sample picked by the inspector, (4) the tests themselves are not perfect, and (5) results of chemical testing generally are not available until well after the meat, poultry, or egg product and the remainder of the production lot have entered the distribution network and have been consumed.

Immediate and substantial increases in testing of the food supply are needed. Such testing must be on a sample basis but at a level of effort sufficient to identify threats that are less than nationwide in scope, and they must be rapid and consistent with what is known about the distribution and development of food hazards. In this context, automated laboratory operations provide an attractive approach. There is also a need to improve the use of sampling to protect food safety. Sample size alone is not enough to provide usable information; how the sample is designed and selected is even more important.

NEW APPROACHES TO TESTING

Other reports in this volume describe the advantages of new high-throughput approaches to testing. Despite these advantages, there are some reasons for concern about relying too greatly on centralized, rapid, high-volume analytical facilities. First, consolidating testing and reducing the number of test facilities will ensure that the impact of any mistakes that occur will increase. Although serious laboratory problems will become less frequent, when a mistake is made, it will be immense. Second, several statistical issues require careful consideration. The effects of biases and correlated errors may or may not be reduced, but they are less likely to be detected when there are fewer players on the scene and are much less likely to be considered in interpretations of test findings. In addition, agency heads and administrators responsible for regulation do not like to think about problems in the data that they receive, and if there is little or no opportunity to detect and assess any problems, far more weight may be given to scientific data than is warranted.

Although new chemical analytical procedures may be useful, they should be implemented in ways that meet the same criteria applied to any other innovation. First, as noted, larger and more representative samples are needed to increase the likelihood of detecting noxious agents, whether chemical, biological, or physical. There will be a need to consider what size of effect (perhaps after translation into what size of exposure) we want to be reasonably sure to detect through monitoring. The present chemical testing program of the U.S. Department of Agriculture can, in rough terms, detect with 95 percent assurance a 1 percent or greater level of risk, if it is uniform over the course of a year and acts on the entire nation, provided that the outcome does not occur at all in the absence of a new hazard. The likelihood of detection diminishes rapidly as the "natural" incidence of the hazard increases from 0 and as the region of concern shrinks to smaller agricultural areas or shorter time periods.

In the event of a major food-related problem, whether natural or deliberate, a further and massive short-term increase in the need for data will occur. These data needs will require facilities and skills that cannot be produced overnight. Furthermore, even interpretations of the data must be "on the shelf" in order to save the time needed to centralize, process, and study the data. While there is still time for reflection, it is important to determine ways in which to react to specific possible threats.

More and better methods for testing food are needed to assure that our nutrition is optimal. Such tests will generally focus on gross elements of diet (carbohydrate, protein, and fat) plus vitamins and minerals, rather than on trace elements and residues. However, it may be that the needs and methods for testing to improve nutrition will differ from those that are needed for safety testing.

How can rapid, high-volume, inexpensive testing help to protect the U.S. food supply? Such testing may improve:

- Research in basic food science, where obtaining reliable data is often the most difficult and time-consuming step in the process.
- The safety of new technologies (e.g., residues of antibiotics and hormones, irradiation of foods, genetic engineering).
- Testing of the stream of foods to detect and remove contaminated foodstuffs.

- Sample testing to characterize, monitor, and ultimately eliminate hazards by identifying feasible means to deal with problems at an early point in food production, processing, and distribution.

The distinction between testing the stream of foods to divert contaminated products and testing to prevent future mishaps is critical. Much of our inspection and testing is still founded on after-the-fact detection and diversion.

Although recent developments can vastly increase the range, depth, and speed of data generation, they can do little for other steps in food testing, such as selecting and collecting proper samples or transporting them to a central laboratory. When the cost or time needed to process a sample is the limiting factor, large-scale, high-throughput laboratories may help. Otherwise, they are likely to contribute little, except perhaps by modestly reducing the unit costs of data generation.

TERRORISM

Bioterrorism makes many of us uneasy. However, the reality is that the highly distributed production, transportation, and distribution systems for food in the United States may make interruption of the food supply an unattractive target for terrorists, although numerous smaller acts could sow fear, confusion, doubts about the civil authorities, and general unrest.

What might a terrorist view as opportunities related to our food supply? Major damage to a great range of potential targets—humans, animals, plants, and even whole ecosystems—could advance terrorist goals, even if the damage is localized to the level of a town or neighborhood.

One grave possibility is the modification of biological or other agents that could damage substantial parts of the food economy. Such agents may be produced in ways that are relatively fast, inexpensive, and easily concealed and that do not require vast knowledge or technical skill. Further, such agents may be easy to transport and distribute. Microbiological agents might be especially attractive for terrorism because they reproduce themselves and can be made in ton lots. However, they may spread beyond the boundaries of the target area, and a terrorist organization might not want to use an agent that could affect its own community. On the other hand, it seems unlikely that a terrorist could have access to a biological, chemical, or physical food hazard the outcomes of which lie far outside already known risks. Rather, the terrorist is likely to be the vector of a known agent (possibly modified).

The details of some of the most important questions about our future food supply, including vulnerability to terrorism and appropriate methods for dealing with these changes, are not yet clear. The combination of unclear problems and unclear methods is worrisome. At present, understanding of the uncertainties and risks of terrorist attacks on the food supply lies mostly in the realms of laboratory research, conflict resolution, and pure guesswork, although the prevention or control of specific problems is more likely to fall within the realm of epidemiology, engineering, technology, and agricultural science. All of these matters are essentially highly multidisciplinary.

What can be done to reduce the likelihood and scope of terrorist attacks on our food supply? We will need the following:

- Complete, accurate, and timely knowledge about what threats exist or may be created.
- Very early pinpoint identification of terrorist acts.
- Adequate means of defense against those threats and swift deployment of countermeasures.
- Capabilities for major transport of foodstuffs into an affected region.

Overall, bioterrorism focused on multiple small areas using microbiological agents is a credible threat now and in the future. It would not be surprising to learn that some efforts have already been launched.

FEDERAL APPROACHES TO FOOD SAFETY

Our federal structure for ensuring the safety and integrity of the food supply is byzantine. At least a dozen federal agencies have major responsibilities for food safety, and many others have important but smaller roles. These agencies administer more than 35 separate statutes, and they have developed more than 70 Memoranda of Understanding to deal with problems that cross agency boundaries. In Congress, 28 separate House and Senate committees are responsible for oversight of these food safety activities (National Research Council, 1998). The Food Safety and Inspection Service inspects meat and poultry. The Center for Food Safety and Applied Nutrition of the Food and Drug Administration (FDA) inspects fish but takes no regulatory action. Other agencies are responsible for the inspection of fruits and vegetables, but they rarely resort to using their regulatory powers. Testing is carried out and standards are set by the FDA and the Environmental Protection Agency. Imports are handled in a range of different ways. In addition to a more effective organizational structure, lacking at the federal level are:

- Up-to-date knowledge about possible agents and their effects, including exposure, biological mechanisms, and treatment or prevention.

• The capacity for extremely rapid response to indications of a new problem.

• The data needed for routine operations.

• Preparations for an effective response, including public education, stockpiling of the materials most likely to be needed, and improved linkages to make use of the substantial capacity of state and local governments in supporting an effective response.

Several approaches for response have been proposed, including forming a food safety council (of agency heads) and appointing a food safety "czar" (a coordinator without the staff or authority to direct needed action).

A council of agency chiefs would not be able to deal rapidly and effectively with the full range of possible microbiological, chemical, and physical hazards, as well as with the integrity of the food supply itself. The present structure is simply too fragmented to deal with the hazards outlined above in an effective and coordinated manner. What are the alternatives? The notion that a council of elders (e.g., agency heads) could deal with a significant problem promptly and decisively is unlikely. The notion that a czar could accomplish the task through persuasion is even more improbable. Either structure seems to ensure inadequate response to any major broad threat to the U.S. food supply. This makes a compelling case for the unification of responsibilities for food safety, whether microbiological, chemical, or physical. Our country needs a single independent food safety agency.

A Committee of the National Academy of Sciences recently made the following recommendation:

> To implement a science-based system, Congress should establish by statute a unified and central framework for managing federal food safety programs, one that is headed by a single official and which has the responsibility and control of resources for all federal food safety activities, including outbreak management, standard-setting, inspection, monitoring, surveillance, risk assessment, enforcement, and education. (National Research Council, 1998)

This committee developed its recommendation based on the hazards that are already with us, paying almost no attention to bioterrorism or the other significant problems mentioned elsewhere in this paper. When bioterrorism is added to the mix, the case for prompt and sweeping change becomes compelling. While additional tinkering with the details of our food safety system might be helpful, the consolidation of responsibilities, authorities, and resources for food safety into a single high-level agency is critical.

REFERENCES

Lowrance, W. W. 1976. Of Acceptable Risk: Science and the Determination of Safety. Los Altos, Calif: W. Kaufmann.

National Research Council. 1998. Ensuring Safe Food: From Production to Consumption. Washington, D.C.: National Academy Press.

Stanlis, P. J., ed. 1968. Selected Writings and Speeches by Edmund Burke. Gloucester, Mass.: P. Smith.

13

Foodborne Pathogen and Toxin Diagnostics: Current Methods and Needs Assessment from Surveillance, Outbreak Response, and Bioterrorism Preparedness Perspectives[1]

Susan E. Maslanka, Gerald Zirnstein, Jeremy Sobel, and Bala Swaminathan

The consumption of contaminated food causes some 76 million illnesses, 325,000 hospitalizations, and 5,000 deaths in the United States each year. More than 200 known diseases are transmitted through foods; however, in only 18 percent of foodborne illnesses is an agent identified. Current laboratory methods used to identify etiologic agents are relatively slow, since they usually depend on the growth of an enrichment culture and subsequent characterization of an isolate by standard biochemical tests. While the combination of culture, isolation, and biochemical characterization of an isolate is the gold standard for identification of an etiologic agent, newer methods are needed for rapid detection so that prevention strategies can be implemented quickly to reduce the incidence of disease. Many immunoassay-based tests and/or systems show promise in their ability to correctly identify an agent quickly, particularly when the tests are used in conjunction with methods that concentrate the target agent. Some of these new technologies incorporate unique detection devices, such as biosensors, into immunoassay tests to increase the sensitivity of detection. Others, such as DNA chip technologies, have the capacity to screen rapidly and simultaneously for many infectious disease agents. Finally, a new technology may be on the horizon to identify a specific etiologic agent by assessing the change in expression of host immune response genes. This could make it possible to use a single universal testing protocol to determine the agent of disease.

---

[1]Use of trade names is for identification only and does not imply endorsement by the Public Health Service or the U.S. Department of Health and Human Services.

Consumers increasingly demand high-quality, readily available, and safer food and water supplies. The United States has made a major commitment toward improving the nation's food supply through the interagency National Food Safety Initiative established in 1997. The establishment of this initiative resulted in a cooperative effort by federal and state partners to provide protection to consumers from contaminated food. Water will be included as a component of the initiative in 2001 by the addition of the Environmental Protection Agency to the federal partnership. The scope of these coordinated efforts needs to be broadened to include the development, validation, and deployment of sensitive, rapid methods for agent detection in investigations of intentional and unintentional contamination of our food and water supplies.

BURDEN OF FOODBORNE ILLNESS IN THE UNITED STATES

The spectrum of illnesses caused by consumption of contaminated foods may range from self-limiting mild gastroenteritis to life-threatening neurological, hepatic, and renal syndromes. Mead et al. (1999) have estimated the number of illnesses, hospitalizations and deaths in the United States using data from various national surveillance systems. Their estimates indicate that contaminated foods cause approximately 76 million illnesses, 325,000 hospitalizations, and 5,000 deaths in the United States each year. The economic burden is estimated to be $9 billion to $32 billion. More than 200 known diseases are transmitted through foods; the agents of foodborne illnesses include viruses, bacteria and their toxins, fungi and their toxins, parasites, poisonous plant components, marine biotoxins, heavy metals, and possibly prions. However, in 82 percent of foodborne illnesses, the identity of the pathogen is unknown. Of 1,500 deaths each year due to known pathogens, 75 percent are caused by *Salmonella, Listeria monocytogenes*, and *Toxoplasma*.

CHANGING EPIDEMIOLOGY OF FOODBORNE DISEASES

The epidemiology of foodborne diseases has undergone profound change in the past two decades. Some factors influencing this change are the global distribution of food supplies to meet increasing consumer demands for greater diversity of foods, centralization of food production, processing and distribution to improve efficiencies and reduce costs, demographic changes occurring in industrialized nations that have resulted in increases in the proportion of the population with heightened susceptibility to severe foodborne infections, changes in food-related behaviors by consumers, and dramatic increases in world travel (Kaferstein et al., 1997; Swerdlow and Altekruse, 1998). One negative effect of the high-degree consolidation of food production, processing, and distribution is that food-safety-related failures may affect large numbers of people over large geographic areas and may have disastrous public health consequences. Because of the explosive increases in international travel, new and emerging pathogens from one corner of the world are able to arrive at a location thousands of miles away in a matter of hours. Transcontinental flights themselves offer many opportunities for transmission of foodborne disease (Tauxe et al., 1987). In addition, the manufacturers and/or distributors of a contaminated food are likely to encounter dire financial and public relations consequences following the implication of their products as a source of widespread illness.

Examples of large-scale (several thousands of cases) foodborne outbreaks are listed in Table 13.1. The 1985 outbreak of *Salmonella* ser. Typhimurium infections was most likely caused by improper switching of the stainless steel pipes in the milk-processing facility, which resulted in raw milk coming in contact with pasteurized milk (Ryan et al., 1987). Interestingly, the outbreak was first recognized as a potentially large one when clinical laboratories in the region ran out of laboratory supplies for culturing *Salmonella* from ill persons. The ice-cream-associated outbreak of *Salmonella enteritidis* infections in 1994 was caused by improper cleaning and sanitation of the ice cream premix tanker that was used previously to transport raw liquid eggs (Hennessy et al., 1996). The Japanese outbreak of *Escherichia coli* O157:H7 infections was most likely caused by contamination of seeds used for sprouting or contamination of water used in the sprouting process (Michino et al., 1999).

Some examples of large waterborne disease outbreaks are listed in Table 13.2. In the largest reported outbreak of typhoid fever in India in

TABLE 13.1 Examples of Foodborne Disease Outbreaks That Have Affected Large Numbers of People

Year	Location	Etiologic Agent	Food Vehicle	Number of Persons Affected
1985	Midwestern U.S.	*Salmonella* serotype Typhimurium	2% pasteurized milk produced by a large dairy	250,000
1994	Nationwide, U.S.	*Salmonella* ser. Enteritidis	Ice cream	224,000
1997	Sakai city, Japan	*E. coli* O157:H7	School lunch, radish sprouts	10,000

TABLE 13.2 Examples of Waterborne Disease Outbreaks That Have Affected Large Numbers of People

Year	Location	Etiologic Agent	Source of Infection	Number of Persons Affected
1966	Riverside, California	*Salmonella* ser. Typhimurium	City water supply	16,000
1975-1976	Sangli town, Maharashtra state, India	*Salmonella* ser. Typhimurium	Sewage contamination of well water	9,000
1992	Kanpur, India	Hepatitis E	Inadequate chlorination	79,000
1993	Milwaukee, Wisconsin	*Cryptosporidium parvum*	Contaminated municipal water	>400,000
1999	New York City	*E. coli* O157:H7 and *Camplylobacter* spp.	Contaminated well water	900

1975 to 1976, 6.7 percent of the towns population was affected. The problem was caused by overflow of sewage into wells that were the sources of drinking water for the town. A large outbreak of hepatitis E infections occurred in Kanpur, India, in 1992, which was probably caused by fecal contamination of river water and inadequate chlorination. Waterborne outbreaks are not limited to areas with inadequate public health systems. In 1993 the Milwaukee outbreak of cryptosporidiosis rapidly depleted the available supply of over-the-counter antidiarrheal medications in that area and overwhelmed the health care system (Colley, 1995). In June 1978, 19 percent of the population of Bennington, Vermont, was affected by diarrheal illness caused by *Campylobacter jejuni* contamination of the town's drinking water source (Vogt et al., 1982). Despite the recent enhancement of conventional and laboratory-based surveillance for foodborne bacterial diseases in the United States, a 1999 waterborne outbreak of gastrointestinal illness affected more than 900 persons who attended a county fair in New York state; 122 *E. coli* O157:H7 infections and 51 *Campylobacter* spp. infections were culture confirmed. Eleven persons developed hemolytic uremic syndrome and two died (Centers for Disease Control and Prevention, 1999).

Large outbreaks like those described above require enormous investments of federal, state, and local resources to investigate and control. These resources are needed for epidemiologic identification of the source of the disease and subsequent laboratory isolation, identification, and characterization of the etiologic agents. The National Food Safety Initiative (NFSI) provides a funding mechanism to improve foodborne disease surveillance and outbreak investigations (U.S. Department of Health and Human Services, Department of Agriculture, and Environmental Protection Agency, 1997). Key components of NFSI include enhanced foodborne disease surveillance, improved responses to foodborne disease outbreaks, improved risk assessment, development of new research methods, improved food safety inspection systems, and improved food safety education.

POTENTIAL OF FOODBORNE AGENTS AS MECHANISMS FOR BIOTERRORISM

Intentional contamination of our food and water supplies is a real threat. There have been two well-documented instances of intentional contamination of foods with pathogenic microorganisms. In 1984, members of a religious commune in Oregon attempted to influence the outcome of a local election by intentionally contaminating salad bars in several restaurants with *Salmonella* ser. Typhimurium. The outbreak affected at least 750 persons, and *S.* Typhimurium was cultured from stool specimens of 388 persons (Török et al., 1997). In 1996, 12 of 45 laboratory workers at a large medical center in Texas became infected with *Shigella dysenteriae* type 2; the outbreak was associated with eating pastries or doughnuts that had been placed in the staff break room on a specific day. Epidemiologic and laboratory investigations strongly suggested intentional contamination of pastries by someone who had access to the bacterial stock cultures in the medical center's laboratory and who was familiar with the methods of culturing the bacteria (Kolavic et al., 1997).

Unlike some potential threat agents (i.e., smallpox) for which the sources are limited, many foodborne agents such as *Salmonella, E. coli* O157, and even botulinum toxin, are relatively easy to obtain or produce. Many of the agents are stable under a variety of conditions and so could easily be added to food and water supplies before consumption. Although there was no reason to suspect foul play in any of the three foodborne outbreaks listed in Table 13.1, each could have easily been caused by intentional contamination by one or more persons involved in some way in food processing, preparation, or transport. As a vehicle for intentional dissemination of pathogens or toxins, water is particularly worrisome because it is an extremely efficient vehicle that could incapacitate large numbers of people within a very narrow window of time. Although none of the outbreaks listed in Table 13.2 were due to intentional contamination of the water supply, the data in this table illustrate the magnitude of problems that could be created by intentional acts of bioterrorism in which water supply is used as the vehicle for contamination.

Some foodborne disease agents require only a small inoculum to cause disease. Shigellosis can be caused by just a few hundred organisms; the infective dose of *E. coli* O157:H7 is thought to be even less (Hornick, 1998). Botulinum toxin is one of the most potent toxins known; it has been estimated that 1 gram of botulinum toxin is enough to kill 1.5 million people (McNally et al., 1994). Introduction of botulinum toxin into a food source would severely strain the resources of the health care system (e.g., antitoxin, hospital support, mechanical ventilators). Although perhaps less deadly, other pathogens intentionally introduced into food and/or water supplies could also negatively affect the ability of a community to respond to the disease. Widespread disease could easily overburden the health care system (hospitals, doctors, medical supplies), the public health system (epidemiologists, diagnostic testing laboratories), and emergency response teams (police, paramedics, decontamination crews). In addition, lack of consumer confidence in the quality of the food and water supplies would be an additional burden on community governments. Capacity for early detection of intentional contamination of the nation's food and water supplies is vital to minimize the impact on community health.

DESCRIPTION OF DIAGNOSTIC NEEDS AND LIMITATIONS OF CURRENT METHODS

There is a great deal of overlap between laboratory support for surveillance for foodborne (or waterborne) diseases, outbreak investigations, and bioterrorism preparedness. The methods used for all these public health response activities must be sensitive and specific in order to accurately identify cases. Ideally, the diagnostic method used for any of these activities also would be rapid, easy to use, and cost effective. In reality, some of these additional requirements are sacrificed to support the most important purpose of a particular activity while maintaining accuracy in case identification. In general, the requirements of a method are dictated either by a need to respond rapidly to an event or to perform extensive characterization of isolates. Table 13.3 lists some of the additional attributes that must be considered, other than sensitivity and specificity, which are critical for all activities, when evaluating a laboratory method for a specific purpose.

Laboratory-based surveillance activities aim to capture all confirmed cases of a given disease. As centralized production and wide distribution

TABLE 13.3 Public Health Response Method Requirements[a]

Public Health Response	Speed	Ease of Use	Cost Constraints	Strain Discrimination
Surveillance	Low	Medium	Low	High
Outbreak response	High	Medium	Medium	Only if relevant to treatment [b]
Bioterrorism event	High	High	High	Only if relevant to treatment [b]

[a]It is assumed that sensitivity and specificity are critical method requirements for all types of public health response.

[b]Although immediate response may not require strain discrimination, subsequent investigations to identify the source of disease require methods that are highly discriminatory.

of food become more common, foodborne disease outbreaks increasingly occur over dispersed areas. Identifying increases in specific subtypes of pathogens through ongoing comparison of surveillance data with historical baselines may be the only way to detect outbreaks among large numbers of persons that lack a clear geographic focus. A good example of this is the ice-cream-associated *Salmonella* outbreak of 1994 (Hennessy et al., 1996). Because surveillance, by its nature, entails capturing the universe of all cases in a population, it requires the processing of high volumes of samples. Surveillance data continue to be used to determine long-term trends in pathogen incidence. However, incidence is documented by sporadic case counts as well as through acute outbreaks (Centers for Disease Control and Prevention, 2000; Glynn et al., 1998).

Laboratory-based outbreak response activities aim to rapidly identify the disease agent. The chosen method should be capable of identifying the suspected pathogen rapidly so that investigations can occur, food vehicles can be identified, and intervention strategies can be implemented to limit the number of cases of disease. Next in importance are ease of use and cost. The tests are usually performed by highly skilled laboratory workers and are more likely to be conducted in state and federal laboratories. Unlike surveillance, identification of the pathogen in a foodborne outbreak may not require testing large numbers of specimens. The majority of unintentional foodborne outbreaks are due to a single pathogen, and confirmation of a single pathogen in 10 to 20 epidemiologically linked patients with a compatible clinical syndrome is usually sufficient. Subtyping of pathogens may be important for epidemiologic investigations of outbreaks, especially when relatively high background rates of infection occur. Identification of a subtype associated with the outbreak allows exclusion of nonoutbreak cases from analytical studies, thereby increasing the power of such studies to implicate a food vehicle.

Laboratory diagnostic response to bioterrorism, as in unintentional outbreak response, also aims to rapidly identify the disease agent. The constraints on appropriate methods for response to a bioterrorism event are great. The method(s) must be rapid to help minimize the potentially large number of cases, and it must be easy to use in order to be effective. The most appropriate method potentially could be used by minimally trained first responders to the event. The need to have a large number of tests available at local sites throughout the country mandates that the method be highly cost effective. To increase the burden of illness or to confound investigators, a bioterrorist might attempt to contaminate food(s) with more than one pathogen. This means that additional tests with different methods may be required even after identifying one agent. Nevertheless, it is appropriate to test a limited number of samples, as in outbreak response to unintentional contamination, with the option of testing a representative sample of cases occurring subsequent to identification of an initial pathogen to rule out "second waves" of contamination with other pathogens. Similar to outbreak investigations is the need for discriminatory methods that could provide data to trace the event back to the source and would provide law enforcement personnel with information to link the event to a perpetrator(s).

The Centers for Disease Control and Prevention, in collaboration with the Association of Public Health Laboratories, the Department of Defense, and other federal partners, has established a Laboratory Response Network for Biological Terrorism. The U.S. Congress allocated funds to establish this network in 1998, which links response to bioterrorism events by various laboratories (local, state, federal). The intent of the network is to provide standardized protocols to appropriate testing laboratories that are as close as possible to the geographic area of the release of the biological agent. Laboratories are given designations (Level A, B, C, or D) based on their capacity to handle a specific agent and perform specific diagnostic tests. Some agents and/or protocols that can be handled without special safety equipment, such as a biological safety cabinet, can be processed and identified at the lowest level (Level A) in a hospital or clinical microbiology laboratory. Other agents or protocols, because of their particular hazardous status, must be handled at a federal laboratory level (Level D). With such a network, laboratory response time to an announced or unannounced bioterrorism event should be reduced because emergency response personnel know the appropriate mechanisms for submitting specimens for testing.

PROMISING NEW TECHNOLOGIES FOR RAPID DETECTION OF FOODBORNE PATHOGENS AND THEIR TOXINS

Microtiter Immunoassay-Based Detection Methods

There are a number of different immunologic technologies that fit the requirements of clinical laboratories. Although some are more laborious than others, each has a definite role in analysis of pathogens and toxins likely to be encountered in outbreak investigations and in response to bioterrorist events. Many of these technologies detect only a few pathogens or toxins and need to be adapted for the detection of multiple foodborne/waterborne agents.

The majority of immunological test formats for rapid automated analysis of clinical and food samples are based on the noncompetitive "sandwich" enzyme linked immunosorbent assay (ELISA). This method is generally favorable for most clinical and food samples because materials present in the sample, including proteases and noncompetitive enzyme inhibitors that could substantially alter enzyme activity, are removed in the washing step prior to detection (Swaminathan et al., 1985; Clark and Engvall, 1985). Both the Alert system (Neogen, Lansing, Mich.) and the RIDASCREEN (R-Biopharm GmbH, Darmstadt, Germany) are sandwich ELISAs that use antibody-coated microtiter wells but different enzyme/substrate detection systems. Both can detect the presence of *Staphylococcus aureus* enterotoxins. However, the RIDASCREEN microtiter strips can differentiate between toxins A, B, C, D, and E.

The Eia Foss system (Foss North America, Eden Prairie, Minn.) also utilizes sandwich ELISA technology but with fluorescence detection. However, the Eia Foss system is unique in that it combines automated ELISA and immunomagnetic separation (IMS) methods in one system. The Eia Foss system can accommodate up to 27 samples per run and can complete the tests in less than 2 hours. Immunomagnetic separation improves the sensitivity and specificity of the test.

Two examples of fully automated microplate and sample handling systems are the Transia Elisamatic II (Diffchamb AB, Gothenburg, Sweden) and the TECRA MINILYSER processor (TECRA International, Willoughby, NSW, Australia). With these systems, results are available within approximately 3 hours, including the sample preparation and the ELISA analysis. Both can detect staphylococcal enterotoxins (SET). TECRA also produces a TECRA SET ID VIA kit for identification of individual toxins in samples found to be positive with the TECRA SET VIA kit, providing the same valuable advantages as the RIDASCREEN product with toxin-specific identification capability but with the option of fully automating the assay.

The VIDAS (Vitek Immuno Diagnostic Assay System) and the mini VIDAS (bioMerieux, Hazelwood, Mo.) also are automated systems that have some unique capabilities not found in other products. A reagent strip with plastic wells containing all of the individual reagents and washes needed for a particular immunoassay is punctured during the course of testing by a solid-phase receptacle. The receptacle is coated on its inside surface with antibody directed toward the target antigen in the sample. Additionally, the receptacle serves as a pipette tip to transfer sample and reagents during the assay after puncturing the foil covering protecting the contents of the individual wells. The VIDAS module, which can run 30 tests at a time, is controlled by a computer module that executes all of the robotic procedures. As many as four VIDAS modules can be operated by a single computer module, for a total capacity of 120 simultaneous tests. A unique advantage of the VIDAS module over other forms of automated immunological testing instruments is that the 30-sample capacity of the module is divided among five chambers with six slots for reagent strips in each chamber. Each chamber can be programmed to run assays with incubation temperatures, hold times, and reagents completely different from the conditions used to perform assays in an adjacent chamber. This gives the VIDAS system a significant degree of flexibility when testing simultaneously for multiple analytes.

Rapid, Immunoassay-Based Kits

Commercial immunological assays for mycotoxin detection generally rely on rapid detection methods because of the need to test agricultural commodities for toxins and quickly release the product for further processing or shipment. Agri-Screen, Veratox, and Veratox HS (Neogen, Lansing, Mich.) are rapid tests for mycotoxins that use a competitive direct ELISA. The assay is based on competition between free toxin in the sample and the control toxin (enzyme-labeled conjugate) for the antibody binding sites. Testing can be completed in 5 to 10 minutes using the Agri-Screen and Veratox products, while the Veratox HS product, which has a higher level of sensitivity (< 2 ppb), takes 20 minutes to perform. Agri-Screen is sold as a field kit and is a qualitative test that allows a sample to be scored as positive or negative for the analyte by visual comparison with known controls; therefore, it does not require the use of laboratory equipment. Veratox and Veratox HS require the use of microwell readers. Several companies have developed immunoassay cards for pathogens and toxins that are similar in design to pregnancy test kits. The VIP test (BioControl Systems, Bellevue, Wash.), the Transia Card (Diffchamb AB, Gothenburg, Sweden), the REVEAL test (Neogen, Lansing, Mich.), and the *Listeria* Rapid Test and the *Clostridium difficile* Toxin A test (both by

Oxoid, Basingstoke, Hampshire, England) are immunoassay cards based on the sandwich ELISA method. All of the immunoassay cards rely on the formation of an antigen-antibody complex, as the test sample (usually 100 to 135 µL) is pulled by capillary action through the test card. The antigen-antibody complex becomes trapped on an immobilized line of antibody. The line becomes visible because of the presence of a dye, chromogen, or colloidal gold bound to the antigen-antibody complex. Nothing is required of the operator except the preparation and loading of the sample to be tested. Results can be read within 5 to 20 minutes, depending on the construction method for the card.

The rapid immunological tests are easy to perform, relatively inexpensive, reasonably fast, and able to be performed immediately upon receipt of a sample rather than after all the samples are run in a large batch. Therefore, they are particularly well suited for field use during outbreak investigations or in first tier response laboratories in the bioterrorism response network.

Biosensors

A biosensor can be thought of as a biorecognition assay or "sensor" in close proximity to or linked to a signal receptor system or "transducer" (Goldschmidt, 1999). When the sensor reacts with its target, the transducer records the changes in intensity that are related to the concentration and/or activity that occurs. Sensors can be made to detect microorganisms, tissue culture cells, and a variety of biomolecules such as enzymes, antibodies, antigens, DNA, and RNA. The technology behind the operation of the transducer may be based on electrochemical/electric, optical, thermal, or other properties, and most instruments can be automated by interfacing them with a computer and recording device.

One of the newer immunology-based detection systems is the BIACORE system (Biacore International AB, Uppsala, Sweden). This is a microchip-based system that permits the detection of biomolecules and the association/dissociation kinetics between macromolecular complexes. The detection principle of the BIACORE microchip is based on surface plasmon resonance, which allows the measurement of changes in the refractive index at the sensor chip surface as they occur. With one reactant immobilized on the sensor chip, a second reactant can be injected over the chip's surface. Changes in mass at the chip surface due to the interaction of reactants alter the refractive index and affect the resonance angle. A variety of assays are possible with this technology, including a direct assay (an antibody bound to the sensor chip surface captures an antigen), sandwich assays similar to ELISA methods, and inhibition assays (sensor-bound antibody competes with antibody added to the sample; Robinson,

1997). In addition, a method for thermodynamic analysis of biomolecular interactions using this technology has been described (Roos et al., 1998). Among other applications, the BIACORE system has been used to examine the interactions between antibodies and antigens (Malmqvist, 1993) and to detect *Salmonella* (Haines and Patel, 1995). A major pharmaceutical company is adopting biosensor assays using the BIACORE 2000 to support preclinical and clinical trials for the detection of antibodies. One noted advantage of the biosensor assay is increased throughput, since detection is label free and occurs in real time. Also, direct detection of binding interactions may enable assessment of antibody avidity (Swanson et al., 1999).

Another fully automated biosensor system based on optical detection is the IAsys Auto+ Advantage (Affinity Sensors, Cambridge, United Kingdom). The IAsys system utilizes resonant mirror (also referred to as evanescent wave) technology, which can detect changes in refractive index (or mass) due to binding of molecules at the surface of a biosensor cuvette. Changes in mass (binding) result in changes in the resonant angle, which are linear with respect to mass. This system uses an open cuvette design, which is intended to minimize sample contamination and aid sample recovery. This technology has been used successfully to detect bacterial toxins in food samples by utilizing a sandwich biosensor with two antibodies. Little or no background interference was noted, even when detecting toxin in complex food matrices (Rasooly and Rasooly, 1999).

The use of biosensors for analysis of clinical and food samples for surveillance, outbreak investigations, and response to bioterrorism events is most certainly in its infancy. In the future, the use of low-cost biosensors combined with efficient telemetry should enable rapid monitoring of environmental contamination in theaters of war and during political and social events with increased risks of terrorist attack.

Microarrays

With the invention of microarray (DNA chip) technology, researchers who used to spend many hours performing tedious manual experiments to study one gene at a time (Zirnstein et al., 1999; Mustapha et al., 1995; Heruth et al., 1993) will be able to examine thousands of genes at a time. As the manufacture and use of DNA chips becomes more common, the price of this type of molecular genetic analysis will become more accessible for routine use by public health laboratories. Nucleotide base pairing (i.e., A-T and G-C for DNA, A-U and G-C for RNA), or hybridization of DNA and RNA molecules, is the scientific principle by which microarrays function. Microarrays generally have nucleic acid probe spot sizes of 200 microns or less in diameter, and these arrays generally have thousands of spots. The "probe" is a tethered or fixed nucleic acid of known sequence, while the "target" is the free nucleic acid in a sample. The primary uses for DNA microarray technology are to identify gene sequences and mutations in gene sequences and to determine the expression level or abundance of genes.

Two general formats of DNA microarray technology are in use. The first format uses probe cDNA (500 to 5,000 bases in length) immobilized on a solid surface, usually glass, using robotic spotting. The array is then hybridized to a set of target nucleic acids, either one at a time or as a mixture of targets. This method is generally referred to as "DNA microarray" (Ekins and Chu, 1999). The second format is an array of oligonucleotide (20- to 25-mer oligos) or peptide nucleic acid probes synthesized in situ on the chip or by conventional synthesis of the oligonucleotide probes followed by immobilization of the probes onto the chip. The array is hybridized to labeled sample DNA, and the identity and abundance of complementary sequences are determined. This second format for microarrays is generally termed DNA chips or GeneChip arrays since they were first developed at Affymetrix, Inc. (Santa Clara, Calif.), a company that sells an integrated GeneChip Instrument System that automates the loading, processing, and analysis of premanufactured GeneChip microarray cartridges. The GeneChip Fluidics station automatically loads the nucleic acid target sample onto the probe microarray cartridge. The fluidics station also controls the delivery of reagents and the timing and temperature for hybridization of the nucleic acid target to the probe array. As many as four probe arrays can be processed independently at one time. Once the hybridization step is complete, messages are displayed on the screen of the controlling computer station indicating that the probe array is ready for scanning. Probe arrays are scanned with a Hewlett-Packard Gene Array Scanner (Hewlett-Packard, Palo Alto, Calif.), which uses an argon-ion laser to excite fluorescent molecules incorporated into the nucleic acid target to generate a quantitative hybridization signal. The fluorescent data from thousands of 3-micron spots within the probe cells of the GeneChip result in a high-resolution image of the probe array and are stored in a file. Data are analyzed using GeneChip Analysis Suite software. GeneChip technology has been useful for species identification in the genus *Mycobacterium* (determined from 16S rRNA sequences) and for detecting rifampin resistance (*rpoB* alleles) in *Mycobacterium tuberculosis* (Troesch et al., 1999), and it should be extremely useful for the detection of bacteria in general.

At this time the cost of premanufactured Affymetrix DNA chips is prohibitively high for many applications (approximately $2,000 per chip). However, a GeneMachines microarray printer (GeneMachines, San Carlos, Calif.) is available for those who wish to produce their own DNA microarray slides. Estimated cost is reduced to $30 to $50 per slide.

GeneMachines currently produces a high-performance microarrayer capable of arraying biological samples from standard 96- or 384-well microwell plates onto a variety of substrates, including glass slides and nylon membranes. Separate hybridization chambers can be purchased to process the arrays. A GenePix 4000A microarray scanner and controlling software (Axon Instruments, Inc., Foster City, Calif.) can be used to collect the fluorescent data generated by the hybridized microarray slides.

The technologies mentioned above rely on passive hybridization of target molecules in solution with the immobilized probe DNAs. Another approach used by Nanogen (San Diego, Calif.) uses electronically mediated active hybridization to move and concentrate target DNA molecules, which accelerates hybridization. Hybridization occurs in minutes rather than the hours required for passive hybridization techniques. This method relies on the fact that negatively charged DNA molecules can be attracted to a positively charged electrode, and if the electrode has DNA probes attached to it that are complementary to the target molecules, the target molecules are "captured." Electronics, in a microchip format, move and concentrate target molecules to one or more test sites on the chip. To remove unbound or nonspecifically bound DNA from each test site, the polarity or charge of the site is reversed to negative, forcing nonspecifically bound and unbound molecules back into solution and away from the capture probes. A dye-labeled reporter binds to a specific target-DNA sequence at the test site, thus allowing the target DNA in the sample to be quantified. This technology can be applied to many types of analyses, including antigen-antibody, enzyme-substrate, cell-receptor, and cell separation techniques. Another advantage to this technique is that the concentration of target molecules over the test site enables a lower concentration of target DNA molecules to be used, resulting in reduced time and labor required for pretest sample preparation.

Flow Cytometry

Flow cytometry was developed in the 1970s for eukaryotic cell analysis. Flow cytometers operate by the movement of particles (cells, beads, nuclei) in single file in an aqueous stream. The cells flow through a focused, high-intensity light beam, which is usually produced by a laser (Raybourne, 1997). Flow cytometry has been used to detect *Listeria monocytogenes* in milk (Donnelly and Baigent, 1986), to distinguish bacterial species with preferential A-T or C-G DNA-binding dyes (Van Dilla et al., 1983), to identify specific species of bacteria using fluorescent in situ hybridization probes for 16S rRNA (Amann et al., 1990), and to differentiate live versus dead cells (Raybourne, 1997). The latter method has been used for antibiotic susceptibility testing and may be applicable to food microbiology or determination of the viability of organisms used in bioterroristic events.

SPECIAL CHALLENGES IN DETECTING PATHOGENIC MICROORGANISMS IN FOODS

One of the problems confronted when trying to analyze food and environmental samples is the need to detect very small numbers of pathogen in a large volume of sample or in a sample that contains large amounts of interfering contaminants. Currently used official methods for pathogen detection in foods, such as contained in the *Bacteriological Analytical Manual* (1996), and similar references, have a minimum detectable limit of one cell per 25-g sample. Improvements in pathogen detection should include approaches that decrease the time required for preenrichment, enrichment, and postenrichment to recover pathogens prior to the application of one of the assays described above. For example, methods currently available for *Salmonella* require a day or two of processing before a rapid diagnostic test can be performed.

Using magnetic beads coated with antibodies is one method that shows potential for reducing the time required for recovery (Holt et al., 1995; Fierens and Huyghebaert, 1996; Mitchell et al., 1994). Antibodies for a specific pathogen are attached to the surface of the superparamagnetic microspheres and effectively capture any pathogen present in the sample. Pathogen-specific magnetic beads have been developed by at least two commercial companies (Dynal AS, Oslo, Norway, and Vicam, Somerville, Mass.). The pathogens captured on the beads can then be cultured or used for further rapid testing by methods such as ELISA or DNA/RNA probes.

A relatively novel approach for the concentration of whole bacterial cells from ground beef, bovine carcass samples, and bovine feces involves the use of hydroxyapatite for the adherence and recovery of viable pathogens (Berry and Siragusa, 1997). Bacterial adhesion to hydroxyapatite is mediated by nonspecific van der Waals and electrostatic attractions (Cowan et al., 1987; Nesbitt et al., 1982). Bacteria, which possess a net negative charge, are adsorbed to hydroxyapatite in a manner similar to the attraction of a negatively charged protein to the positively charged calcium ions of hydroxyapatite (Berry and Siragusa, 1997; Bernardi et al., 1972; Rölla and Melsen, 1975). Kinetic studies revealed that maximum adherence of bacteria to hydroxyapatite takes place within 5 minutes and that both high (10^9) and low (10^3) concentrations of cells yielded comparable percent adherence values within this time.

The use of biotinylated oligonucleotide probes to permit the specific capture of pathogenic DNA in clinical samples containing large amounts of extraneous DNA and inhibitors has also been demonstrated

(Mangiapan et al., 1996). Nonspecific inhibition can be minimized by mechanical disruption and proteinase K digestion. The pathogen-specific DNA that hybridizes with the probes is captured with strepavidin-coated magnetic beads and subsequently amplified using PCR. This approach is reported to be 10 to 100 times more sensitive than procedures in which total DNA is extracted and purified before PCR amplification (Mangiapan et al., 1996). Although sensitivity may be sacrificed, biotinylated oligonucleotides can be directly coupled to the magnetic beads (a direct capture technique), which subsequently decreases the time for identification of the pathogen (Muir et al., 1993).

UNIVERSAL APPROACHES TO SCREEN SPECIMENS FOR INFECTIOUS MICROORGANISMS AND THEIR TOXINS

Current methods for the detection of etiologic agents of foodborne diseases are pathogen/toxin specific. For example, methods for the examination of patient and implicated food specimens in cases of suspected botulism are very different from those in which *E. coli* O157:H7, *Salmonella* sp., or *Listeria monocytogenes* is the suspected etiologic agent. There are significant differences even in the type of specimen(s) collected as well as in the methods for specimen collection and storage protocols for foodborne bacterial and viral pathogens. The variability in agent-specific protocols is likely to significantly delay laboratory identification of the etiologic agent in the event of an unannounced bioterrorism event in which the clinical symptoms alone do not allow investigators to narrow the list of suspected agents to a small manageable number. In the past, rapid screening of patient specimens for immunological response to a broad spectrum of infectious agents has greatly facilitated the identification of a newly emerging etiologic agent; however, because specimens must be processed and screened independently for each potential agent, this process may be labor intensive and time consuming.

This is best illustrated by the approach used in the identification of hantavirus as the etiologic agent of a new respiratory disease syndrome in the southwestern United States (Chapman and Khabbaz, 1994). A variety of clinical specimens were collected during the investigation of this outbreak. These specimens were triaged through several laboratories with expertise in the various potential agents. As one agent was ruled out, specimens were transferred to another laboratory for further testing, until finally the correct identification was obtained. This traditional stepwise protocol used during an outbreak can significantly increase the time to identification of an agent of disease and as a result may hinder the implementation of public health measures to reduce the incidence of disease.

The technologies described earlier may provide some improvement in the rapid diagnosis of foodborne disease; however, assumptions still must be made about the potential agent involved before tests can be performed. It may be possible in the future to characterize and quantify changes in the expression of host genes induced by various etiologic agents using the new microarray technology rather than to directly identify the agent (Manger and Relman, 2000). This technology has the potential to use a universal platform for screening large numbers of clinical specimens for a wide range of etiologic agents within hours or possibly minutes. The advantages would be a uniform protocol for specimen collection, a uniform method for specimen processing and testing, early identification of a suspected agent, and a narrow field of possible agents so that the most effort can be directed toward appropriate specimens and methods for isolation and characterization of a particular agent. This method for agent detection is in its infancy and requires further evaluation to determine the utility in disease diagnostics.

CONCLUSIONS

There has been a recent surge in the development of new technologies for the detection of foodborne pathogens and their toxins. This has been in direct response to the change in the epidemiology of foodborne disease. As public health awareness of new and newly emerging foodborne disease increases, research and development of new methods for agent detection also increase. Although foodborne agent detection has improved dramatically in the past few years, many of the most promising techniques, such as DNA chip technology, are currently too costly to implement effectively on a large scale. However, consumers increasingly demand higher quality and more readily available foods of all types. In addition, they expect protection from disease while consuming these products. Consumers' expectations are fast outgrowing the ability of the public health community to protect them adequately from disease.

As food processing becomes more complex and the diversity of available foods increases, the need for more rapid, sensitive, and sophisticated diagnostic techniques will continue to rise. With a dramatic increase in centralized production and global distribution of foods, outbreaks may affect people in many states or even countries without the geographic clustering that formerly allowed astute clinicians, laboratorians, and public health officials to detect clusters of diseases. The integration of rapid diagnostic tests into laboratory-based public health surveillance is essential to meet this global challenge. In some widespread diffuse outbreaks, cases may be linked and the outbreak recognized only after comparison of agent subtypes with surveillance data available in molecular databases. In addition, the waterborne outbreaks in the United States demonstrate that not only are these techniques needed to help protect consumers of food but that efforts also need to be directed toward the nation's water supplies.

The specter of bioterrorism aimed at food and water supplies, with the ominous possibility of deliberate contamination of foods and water with highly virulent pathogens, further underscores the need for integration of rapid diagnostic capacity into routine surveillance activities, in addition to the parallel rapid-diagnosis infrastructure described in this paper. This is especially important because a bioterrorist attack on the food or water supply may not be readily distinguished from an unintentional event and might well be first detected through routine laboratory surveillance. Further collaboration between government, university, and commercial laboratories is needed to make these promising techniques available for global use. Only then can we meet the increasing demand for even safer food and water supplies.

The United States has made a major commitment toward improving the safety of the nation's food supply by developing and funding the NFSI. This funding was provided to improve foodborne disease surveillance and outbreak investigations, enhance the regulatory control of foods through more frequent inspections, and encourage a prevention-oriented approach to food safety by implementing Hazard Analysis and Critical Control Points (HACCP) programs. The NFSI also requires various federal agencies with responsibility for food safety to coordinate their efforts. This interagency investment in food safety is now beginning to produce results; the addition of efforts specific to water supplies also should provide improvement in consumer protection. Much needs to be done to improve the nation's diagnostic capabilities to achieve a state of readiness to appropriately address bioterroristic events involving either food or water. Our ability to keep pace with rapid technological developments and marketing practices in the food production sector, as well as protect the public from the threats of emerging pathogens and intentional contamination of the food and water supplies, cannot be met without intensive integration of improved rapid diagnostics; enhanced and continually supported public health infrastructure (federal, state, and local agencies), including surveillance and investigative entities; and improved communication components.

REFERENCES

Amann, R. I., B. J. Binder, R. J. Olson, S. W. Chisholm, S. W. Devereux, and D. A. Stahl. 1990. Combination of 16S rRNA-targeted oligonucleotide probes with flow cytometry for analyzing mixed microbial populations. Applied and Environmental Microbiology, 56:1919-1925.

Bernardi, G., M. G. Giro, and C. Gaillard. 1972. Biochimica Biophysica Acta, 278:409-420.

Berry, E., and G. R. Siragusa. 1997. Chromatography of polypeptides and proteins on hydroxyapatite columns: Some new developments. Applied and Environmental Microbiology, 63:4069-4074.

Centers for Disease Control and Prevention. 1999. Morbidity and Mortality Weekly Report, 48:803.

Centers for Disease Control and Prevention. 2000. Morbidity and Mortality Weekly Report, 49:73-79.

Chapman, L. E., and R. F. Khabbaz. 1994. Etiology and epidemiology of the Four Corners hantavirus outbreak. Infectious Agents Disease, 3:234-244.

Clark, B. R., and E. Engvall. 1980. Enzyme Linked Immunosorbent Assay (ELISA): Theoretical and Practical Aspects. Boca Raton, Fla.: CRC Press.

Colley, D. G. 1995. Waterborne cryptosporidiosis threat addressed. Emerging Infectious Diseases, 1:67-68.

Cowan, M. M., K. G. Taylor, and R. J. Doyle. 1987. Energetics of the initial phase of adhesion of Streptococcus sanguis to hydroxylapatite. Journal of Bacteriology, 169:2995-3000.

Donnelly, C. W., and G. J. Baigent. 1986. Method for flow cytometric detection of Listeria monocytogenes in milk. Applied and Environmental Microbiology, 52:689-695.

Ekins, R., and F. W. Chu. 1999. Microarrays: Their origins and applications. Trends in Biotechnology, 17:217-218.

Fierens, H., and A. Huyghebaert. 1996. Screening of Salmonella in naturally contaminated feeds with rapid methods. International Journal of Food Microbiology, 31:301-309.

Food and Drug Administration. 1996. Bacteriological Analytical Manual. Food and Drug Administration.

Glynn, M. K., C. Bopp, W. Dewitt, P. Dabney, M. Mokhtar, and F. J. Angulo. 1998. Emergence of multidrug-resistant Salmonella enterica serotype typhimurium DT104 infections in the United States. New England Journal of Medicine, 338:1333-1338.

Goldschmidt, M. 1999. Pp. 286-289 in The 19th International Workshop on Rapid Methods and Automation in Microbiology, D. Y. C. Fung, ed. Manhattan: Kansas State University.

Haines, J., and P. D. Patel. 1995. Detection of foodborne pathogens using BIA. BIA Journal, 2:31.

Hennessy, T. W., C. W. Hedberg, L. Slutsker, K. E. White, J. M. Besser-Wiek, M. E. Moen, J. Feldman, W. W. Coleman, L. M. Edmonson, K. L. MacDonald, and M. T. Osterholm. 1996. A national outbreak of Salmonella enteritidis infections from ice cream: The investigation team. New England Journal of Medicine, 334:1281-1286.

Heruth, D. P., G. W. Zirnstein, J. F. Bradley, and P. G. Rothberg. 1993. Sodium butyrate causes an increase in the block to transcriptional elongation in the c-myc gene in SW837 rectal carcinoma cells. Journal of Biological Chemistry, 268:20466-20472.

Holt, P. S., R. K. Gast, and C. R. Greene. 1995. Rapid detection of *Salmonella enteritidis* in pooled liquid egg samples using a magnetic bead-ELISA system. Journal of Food Protection, 58:967-972.

Hornick, R. B. 1998. Pp. 736-741 in Infectious Diseases, P. D. Hoeprich, M. C. Jordan, and R. C. Ronald, eds. Philadelphia: J. B. Lippincott.

Kaferstein, F. K., Y. Motarjemi, and D. W. Bettcher. 1997. Foodborne disease control: A transnational challenge. Emerging Infectious Diseases, 3:503-510.

Kolavic, S. A., A. Kimura, S. L. Simons, L. Slutsker, S. Barth, and C. E. Haley. 1997. An outbreak of Shigella dysenteriae type 2 among laboratory workers due to intentional food contamination. Journal of the American Medical Association, 278:396-398.

Malmqvist, M. 1993. Biospecific interaction analysis using biosensor technology. Nature, 361:186-187.

Manger, I. D., and D. A. Relman. 2000. How the host "sees" pathogens: Global gene expression responses to infection. Current Opinion in Immunology, 12:215-218.

Mangiapan, G., M. Vokurka, L. Schouls, J. Cadranel, D. Lecossier, J. van Embden, and A. J. Hance. 1996. Sequence capture-PCR improves detection of mycobacterial DNA in clinical specimens. Journal of Clinical Microbiology, 34:1209-1215.

McNally, R. E., M. B. Morrison, J. E. Berndt, J. E. Fisher, J. T. Bo'Berry, V. Puckett, and N. J. Simini. 1994. Effectiveness of Medical Defense Interventions Against Predicted Battlefield Levels of Botulinum Toxin A. Joppa, Md.: Science Applications International Corp.

Mead, P. S., L. Slutsker, V. Dietz, L. F. McCaig, J. S. Bresee, C. Shapiro, P. M. Griffin, and R. V. Tauxe. 1999. Food-related illness and death in the United States. Emerging Infectious Diseases, 5:607-625.

Michino, H., K. Araki, S. Minami, S. Takaya, N. Sakai, M. Miyazaki, A. Ono, and H. Yanagawa. 1999. Massive outbreak of *Escherichia coli* O157:H7 infection in schoolchildren in Sakai City, Japan, associated with consumption of white radish sprouts. American Journal of Epidemiology, 150:787-796.

Mitchell, B. A., J. A. Milbury, A. B. Brookins, and B. J. Jackson. 1994. Use of immunomagnetic capture on beads to recover *Listeria* from environmental samples. Journal of Food Protection, 57:743-745.

Muir, P., F. Nicholson, M. Jhetam, S. Neogi, and J. E. Banatvala. 1993. Rapid diagnosis of enterovirus infection by magnetic bead extraction and polymerase chain reaction detection of enterovirus RNA in clinical specimens. Journal of Clinical Microbiology, 31:31-38.

Mustapha, A. R., R. W. Hutkins, and G. W. Zirnstein. 1995. Cloning and characterization of the galactokinase gene from *Streptococcus themophilus*. Journal of Dairy Science, 78:989-997.

Nesbitt, W. E., R. J. Doyle, K. G. Taylor, R. H. Staat, and R. R. Arnold. 1982. Positive cooperativity in the binding of Streptococcus sanguis to hydroxylapatite. Infection and Immunity, 35:157-165.

Rasooly, L., and A. Rasooly. 1999. Real time biosensor analysis of staphylococcal enterotoxin A in food. International Journal of Food Microbiology, 49:119-127.

Raybourne, R. B. 1997. Flow Cytometry in Food Microbiology: Detection of Escherichia coli 0157:H7. New York: Marcel Dekker.

Robinson, B. J. 1997. Pp. 87-89 in Food Microbiological Analysis: New Technologies, M. L. Tortorello and S. M. Gendel, eds. New York: Marcel Dekker.

Rölla, G., and B. Melsen. 1975. Desorption of protein and bacteria from hydroxyapatite by fluoride and monofluorophosphate. Caries Research, 9:66-73.

Roos, H., R. Karlsson, H. Nilshans, and A. Persson. 1998. Thermodynamic analysis of protein interactions with biosensor technology. Journal of Molecular Recognition, 11:204-210.

Ryan, C. A., M. K. Nickels, N. T. Hargrett-Bean, M. E. Potter, T. Endo, L. Mayer, C. W. Langkop, C. Gibson, R. C. McDonald, and R. T. Kenney. 1987. Massive outbreak of antimicrobial-resistant salmonellosis traced to pasteurized milk. Journal of the American Medical Association, 258:3269-3274.

Swaminathan, B., J. A. G. Aleixo, and S. A. Minnich. 1985. Enzyme immunoassays for *Salmonella*: One day testing is now a reality. Food Technology, 39:83-89.

Swanson, S. J., S. J. Jacobs, D. Mytych, C. Shah, S. R. Indelicato, and R. W. Bordens. 1999. Applications for the new electrochemiluminescent (ECL) and biosensor technologies. Developments in Biological Standardization, 97:135-147.

Swerdlow, D. L., and S. F. Altekruse. 1998. Pp. 273-294 in Emerging Infections 2, W. M. Scheld, W. A. Craig, and J. M. Hughes, eds. Washington, D.C.: ASM Press.

Tauxe, R. V., M. P. Tormey, L. Mascola, N. Hargrett-Bean, and P. A. Blake. 1987. Salmonellosis outbreak on transatlantic flights; foodborne illness on aircraft: 1947-1984. American Journal of Epidemiology, 125:150-157.

Török, T. J., R. V. Tauxe, R. P. Wise, J. R. Livengood, R. Sokolow, S. Mauvais, K. A. Birkness, M. R. Skeels, J. M. Horan, and L. R. Foster. 1997. A large community outbreak of salmonellosis caused by intentional contamination of restaurant salad bars. Journal of the American Medical Association, 278:389-395.

Troesch, A., H. Nguyen, C. G. Miyada, S. Desvarenne, T. R. Gingeras, P. M. Kaplan, P. Cros, and C. Mabilat. 1999. Mycobacterium species identification and rifampin resistance testing with high-density DNA probe arrays. Journal of Clinical Microbiology, 37:49-55.

U.S. Department of Health and Human Services, Department of Agriculture, and Environmental Protection Agency. 1997. Food Safety from Farm to Table: A National Food Safety Initiative. (A Report to the President.) Washington, D.C.: Department of Health and Human Services, Department of Agriculture, Environmental Protection Agency.

Van Dilla, M. A., R. G. Langlois, D. Pinkel, D. Yajko, and W. K. Hadley. 1983. Bacterial characterization by flow cytometry. Science, 220:620-622.

Vogt, R. L., H. E. Sours, T. Barrett, R. A. Feldman, R. J. Dickinson, and L. Witherell. 1982. Campylobacter enteritis associated with contaminated water. Annals of Internal Medicine, 96:292-296.

Zirnstein, G., Y. Li, B. Swaminathan, and F. Angulo. 1999. Ciprofloxacin resistance in Campylobacter jejuni isolates: Detection of gyrA resistance mutations by mismatch amplification mutation assay PCR and DNA sequence analysis. Journal of Clinical Microbiology, 37:3276-3280.

14

Food Safety: Data Needs for Risk Assessment

Joseph V. Rodricks

INTRODUCTION

Risk assessment entails the systematic organization and analysis of available information and knowledge for the purpose of estimating the likelihood, or probability, that a given activity will in some way harm those engaged in it. Here the term "activity" is used very broadly to encompass all acts undertaken by humans, including the consumption of food.

Several scientific methodologies, none perfect and some highly imperfect, are available to assess risks. For risks that are relatively large and associated with activities that have occurred in the past, it is sometimes possible to obtain direct measures using the tools of epidemiology. At present, the uncountable number of small and even moderately large risks that humans face is generally not susceptible to direct measurement and can be estimated only indirectly. The reliability of a given estimate of risk is a function of the reliability of the information and scientific knowledge on which it is based and is also a function of whether it has been estimated using direct or indirect means. To be useful, a risk assessment should contain a characterization of its reliability and of the uncertainties inherent in it.

Risk assessment, as it is defined here, is essential to an understanding of the safety of food. Too often, and incorrectly, risk assessment has been taken to be activity devoted to the estimation of low-dose risks from carcinogens. Such an overly narrow view of risk assessment is perhaps explained by the fact that the term "risk assessment" was first used in a formal way beginning in the early 1970s in conjunction with its application to carcinogens (National Research Council, 1983). But, as is clear from the definition offered above, any attempt to understand the nature and magnitude of harm associated with the consumption of food, or any of its components or contaminants, falls properly within the compass of risk assessment. The study of food safety is, then, taken to be the quest for an understanding of the relative risks to health associated with the consumption of food or its individual constituents and contaminants, including those of both natural and industrial origin.

Food safety is, however, more than the quest for scientific knowledge about health risks. The term also suggests a social goal—namely, the quest to eliminate or minimize health risks associated with the consumption of food. Knowledge of risks is essential to achieving that goal, but its achievement also requires that appropriate technical and institutional means be available to reduce or eliminate excessive risks. Because a perfectly safe food supply—one in which the consumption of food poses no risks of any kind—is unimaginable (and certainly unknowable even if it were achieved), it is likely that food safety will remain a perennial topic.

Risk assessment relies on information and knowledge developed by the research community and serves to organize and evaluate research findings so that they can be used for social purposes. The links between research, risk assessment, and risk management were first formally described in 1983 by a committee of the National Academy of Sciences/National Research Council (NRC), that was charged with evaluating the development and use of risk-related information in the federal government (National Research Council, 1983, 1994). The committee's recommendations regarding the definitions of and relations among these three activities are summarized in Box 14.1. These definitions and relationships are today widely accepted both in the United States and abroad, although some differences among countries exist, as will be discussed later.

SAFETY

The common-sense definition of a safe activity is one that poses no risk of harm to those engaged in it. In strictly technical terms, such a definition is troublesome because it is never possible to demonstrate the absolute absence of risk. Methods are available to assess the risks associated with various activities, with highly varying degrees of reliability, and to identify the conditions under which risks are likely to be so small that they are of no significant public health concern. A condition of safety is said to have been achieved when a state of sufficiently small risks has been achieved (Rodricks, 1992). There is, of course, continuing debate on

Box 14.1
Risk Assessment Links Research and Policy

Research
- Development of data on hazards and dose-response relationships of agents in the environment.
- Development of data on human exposures to and intakes of those agents.
- Development of basic scientific knowledge pertaining to the above.

Risk Assessment
- Systemic organization of available data to evaluate the likelihood (risk) that human populations will experience any of the hazards associated with environmental agents under their actual or projected conditions of exposure. A four-step process is used (see text).

Risk Management
- Application of criteria to determine whether risk reductions are needed to achieve adequate public health protection.
- Evaluation of options for risk reduction, taking into account law, availability of control technologies, costs, and so forth.
- Decision making.

SOURCE: National Research Council (1983, 1994). "Agents" refers to chemical, biological, or physical substances that may, under some conditions, harm human health. Food is the "environment" that is the subject of this paper, and "agent" refers to any item of the diet or to any constituent or contaminant thereof.

the policy issue of how small risks must be before they can be said to be of no public health concern.

It should be kept in mind that, although the assessment process is conceptually distinct from the risk management process, the two are closely linked because it is essential that the assessment deal with the specific questions that must be resolved in the management process. Thus, for example, if the introduction of some new technology to control microbial risks brings with it risks of a different type, the management process will need to focus on the overall consequences, and this will require that the risk assessment compare and contrast the two risks.

Knowledge that food consumption is essential for life no doubt influences, as it should, almost every risk management decision. With adequate knowledge of both risks and health benefits, it would in theory be possible to ensure that every food safety decision reduces overall risk; risk management options that do not meet this criterion would be eliminated. Given the current state of scientific knowledge regarding risks and benefits, it is not possible to establish such a broad overarching criterion for decision making. Indeed, some of our current laws may even prohibit the application of such a criterion. Nevertheless, the scientific quest for knowledge of risks and benefits will no doubt continue, and this should bring with it increasingly reliable food safety decisions.

CONTENT OF RISK ASSESSMENT

Risk assessment begins with a clear statement regarding the problem to be evaluated. In the context of food safety the typical risk assessment questions are of the following types:

- What are the risks of adverse human health effects associated with the consumption of a new food additive (or pesticide or animal drug residue) under its proposed conditions of use?[1]
- What are the risks of adverse health effects from consumption of food containing a chemical or microbiological contaminant?
- What are the risks of adverse health effects associated with over- or underconsumption of a nutrient substance?
- What are the risks of adverse health effects associated with consumption of a natural nonnutritional constituent of food?

In every case the population of interest also needs to be carefully defined. In most cases it is the general population, but there are circumstances in which specific subpopulations may be singled out for attention.

Although the types and sources of data needed to answer these questions for specific substances will vary, all can be approached in the same way. Once the specific risk assessment question is defined (chemical or microbiological agent and population of concern), the framework for risk assessment set forth in the 1983 NRC study and reinforced in a 1994 report from another NRC committee is applied in the following steps:

Step 1: Hazard Identification. What types of harm or injury (hazards) have been shown (under other circumstances) to be caused by exposure to the agent of interest? Which are likely to be relevant to the population of interest? The information for this step derives from epidemiology and

---

[1]In the case of pesticides especially, occupation and environmental risks also are of concern.

animal studies and from other types of experimental studies bearing on the agent's mechanism of biological action. The goal of this step is to bring together all relevant information, to sort the reliable from the less reliable, and to identify the specific hazards that exposure to the agent might cause in the population of interest, assuming the latter received a sufficient exposure. This step often involves consideration of the problem of applying animal data to humans.

Step 2: Dose-Response Evaluation. For each of the hazards associated with the agent of interest, what is the quantitative relationship between exposure to the agent (dose) and the incidence and severity of harm produced (response)? This information typically derives from the same types of data used to evaluate hazards but often requires that inferences be made regarding the relationship in the region of the dose-response curve that is not susceptible to measurement by any available means (see below, for an explanation of the need for dose-response extrapolation). Note that several different measures of dose may be considered depending on the nature of the available database and the nature of the response. As it is used here, "response" is synonymous with "risk." This step might more properly be termed the "dose-risk evaluation."

Step 3: Human Exposure Assessment. Under what conditions is the population of interest exposed to the agent of interest? "Conditions" refers to the magnitude and duration of the doses received and to information on the variability of exposure within the population. Other factors (e.g., food matrix for pathogens) may need to be included in a description of "conditions." The information regarding human exposure comes from many sources but generally requires knowledge of the concentration of the agent in the relevant items of food and the rates of consumption of those items in the population of interest. The particular measure of exposure (dose) sought in this step must be identical to that identified in Step 2 as the one that is relevant to the risk being assessed.[2]

Step 4: Risk Characterization. The final step of the process involves integration of information and analysis from the first three, and is intended to describe, with appropriate attention to scientific uncertainties, the likelihood that the hazardous properties of the agent of interest will be expressed under the conditions of human exposure. This step thus provides the ultimate answer to the risk question posed.

---

[2]There is sometimes confusion regarding the terms "exposure" and "dose." "Exposure" is generally used to describe human contact with the environmental medium (food, water) containing the agent of concern. "Dose" describes the amount of agent intake per unit of time. Whatever the usage, it is critical that all aspects of exposure that affect risk be considered in Steps 2 and 3.

SOURCES OF FOOD-RELATED RISKS

Food is by far the most chemically complex part of the environment to which humans are directly exposed. There is no reliable estimate of the number of distinct chemical compounds in the different items of food and drink we select for nourishment and pleasure, but it is surely in the hundreds of thousands. The chemical structures of most of these are unknown, but the known constituents display immense variety. To make matters more complex, the chemical composition of the human diet varies from culture to culture and over time within cultures. Food chemists will probably never know more than a small fraction of the chemicals we deliberately put into our mouths every day of our lives (National Research Council, 1996).

The natural constituents of foods and beverages represent the major share of dietary chemicals. In addition to the hundreds of distinct compounds that supply nutritional requirements, there are thousands more that impart flavor and color. Food plants contain large numbers of natural constituents that contribute neither nutritional nor esthetic properties but are present because they play some role in the lives of these plants. It has been estimated, for one small example, that a freshly brewed cup of coffee contains more than 600 distinct (and mostly unidentified) compounds. We also need to note the additional burden of natural products from the hundreds of herbs and spices used in food preparation.

Also among the constituents of the human diet are substances that arise during food and beverage preparation. Fermentation, for example, produces numerous chemical alterations of organic compounds, yielding products bearing little chemical similarity to the starting materials. Little scientific skill, and not much gastronomic skill, is needed to recognize that each variety of wine and cheese possesses a unique chemical composition and that none of these bear much resemblance to grape juice and milk. Roasting, broiling, baking, microwaving, smoking, and other means of preparing and processing foods each sets off dozens of chemical reactions. Because most methods of food preparation have been in use for centuries, people think of the products of preparation as "natural." This is perhaps appropriate, but, strictly speaking, they actually result from human manipulations of raw food products.

Human beings have of course never been satisfied to leave nature as it is and have added substances to food to achieve any number of desirable technical effects. Food preservation using various inorganic salts was probably one of the earliest examples of this practice, but adding substances to color, sweeten, emulsify, flavor, and alter taste perception are also ancient practices that continue to this day.

Many chemicals not directly added to food but that are intentionally used in food production, processing, and storage actually end up in the diet, although usually in very small concentrations. Among these indirectly added substances are residues of drugs and feed additives used in animal production, crop-use pesticides and their metabolites and degradation products, and migrants from materials used in food processing and packaging. Several thousand direct and indirect additives that may be present in foods add a significant increment to the unaccounted numbers of natural substances resulting from food preparation.

Some foods may also contain contaminants—unwanted byproducts of nature or human industry that somehow come to be present in foods. Included are bacterial and fungal metabolites resulting from the growth on food of species of these organisms, organic chemicals of industrial origin, and various metals and other inorganic species that arise either because of their natural presence in soils and water used for food production or because they have locally accumulated to unusually high environmental levels as a result of mining, industrial, or other human activities. For completeness we need to include bacterial metabolites that are not produced directly in food but rather in the intestines following ingestion of foods contaminated with the offending organisms. Some contaminants are regularly occurring constituents of certain foods, whereas others arise only occasionally (and unpredictably) because of a human or natural mishap. As with the other dietary constituents, the total number of possible dietary contaminants is unknown, although the most important ones are fairly well documented. The major categories of food constituents are summarized in Box 14.2.

Box 14.2
The Categories of Food Constituents

I. Natural Components
  A. Nutrients
  B. Nonnutrients
II. Substances Intentionally Introduced (directly and indirectly)
  A. Food and color additives
  B. Substances Generally Recognized as Safe (GRAS)
  C. Veterinary drug residues and feed additives
  D. Pesticide residues
III. Contaminants
  A. Naturally occurring chemicals
  B. Industrial products and byproducts
  C. Pathogenic microorganisms

Risk assessment methods have been developed and applied to all of the food constituent categories listed in Box 14.2. The data used as the basis for the assessments derive from epidemiological, toxicological, and other experimental studies on the agents, as well as from studies regarding the levels of various agents in foods and of the rates and patterns of human food consumption (U.S. Food and Drug Administration, 1994). The various roles of measurement technology in food safety evaluation are discussed in a later section.

RISK-BASED DECISION MAKING

Food safety is assured when risk-based criteria for public health protection are satisfied. Most often the mechanism for achieving this goal involves the steps outlined below:

1. Risks of current or expected human exposures to the agent of interest are assessed.
2. It is determined whether the assessed risks of the agent are excessive, and, if so, risk reduction goals are identified.
3. Estimates are made of the maximum allowable exposure levels for the agent, based on the identified risk reduction goals.
4. Criteria are established, usually in the form of a limit on the allowable agent concentration in foods, so that the maximum allowable exposure levels (Step 3) are not exceeded.

The "limit on concentration," usually termed a "tolerance," becomes the device for determining whether specific lots of food are acceptable from a public health perspective (U.S. Food and Drug Administration, 1994). Thus, tolerances may be established for various chemical or microbial contaminants and for additives and pesticide residues. If they are risk-based tolerances and appropriate risk reduction criteria have been applied, foods containing agents at or below assigned tolerances should be safe to consume.

ROLE OF AGENT MEASUREMENT

Analytical technologies that are known to be reliable and that are reliably applied are essential to ensure a safe food supply. Whatever the agent of concern, there is a need for such analytical technology in at least four areas of food safety:

1. As support for epidemiological, toxicological, and other experimental studies.

2. To determine agent concentrations in foods to support the human exposure assessment step of the predictive risk assessment process.

3. To monitor foods for compliance with risk-based criteria (tolerances).

4. To ensure the successful implementation of industry quality control programs.

Although reliable analytical methods are needed for all four of these activities, it is especially important that automated high-throughput technologies be developed for application in regulatory and public health monitoring and in quality control activities (applications 3 and 4). The numbers of chemical and microbiological contaminants that are potentially the subjects of these applications are large and growing, and rapidity and ease of data access are crucial to prevent the introduction of potentially unsafe lots of food into commerce.

The AOAC International (*www.aoac.org*) is a century-old professional society dedicated to the validation of analytical methods used for food constituents, additives, pesticide residues, and contaminants. The society publishes a volume entitled *Official Methods of Analysis of AOAC International*. Methods published therein have all been subjected to interlaboratory studies to test their reliability, and regulators such as the Food and Drug Administration, the Environmental Protection Agency, and various state agencies rely on AOAC International methods for compliance testing. (Other methods may be used for research purposes.) It is thus critical that those involved in the development of new high-throughput technologies recognize the need for method validation and be familiar with AOAC International procedures.

REFERENCES

National Research Council. 1983. Risk Assessment in the Federal Government: Managing the Process. Washington, D.C.: National Academy Press.

National Research Council. 1994. Science and Judgment in Risk Assessment. Washington, D.C.: National Academy Press.

National Research Council. 1996. Carcinogens and Anticarcinogens in the Human Diet. Washington, D.C.: National Academy Press.

Rodricks, J. V. 1992. Calculated Risks. Cambridge: Cambridge University Press.

U.S. Food and Drug Administration. 1994. Toxicological Principles for the Safety Assessment of Direct Food Additives and Color Additives Used in Food. Washington, D.C.: Center for Food Safety and Applied Nutrition.

PART III

BIOTERRORISM AND BIOWARFARE

15

Biological Weapons: Past, Present, and Future

Ken Alibek

Biological weapons are weapons of mass destruction (or mass casualty weapons, to be precise, because they do not damage nonliving entities) that are based on bacteria, viruses, rickettsiae, fungi, or toxins produced by living organisms. Compared to nuclear, chemical, or conventional weapons, biological weapons are unique in their diversity. Dozens of different agents can be used to make a biological weapon, and each agent will produce a markedly different effect. These differences are shaped by various properties of the particular agent, such as its contagiousness, the length of time after release that it survives in the environment, the dose required to infect a victim, and, of course, the type of disease that the agent produces. Biological weapons are relatively inexpensive and easy to produce. Although the most sophisticated and effective versions require considerable equipment and scientific expertise, primitive versions can be produced in a small area with minimal equipment by someone with limited training.

Biological weapons can be deployed in three ways: by contaminating food or water supplies; by releasing infected vectors, such as mosquitoes or fleas; or by creating an aerosol cloud to be inhaled by the victims. Because industrialized countries have adequate water purification systems, contamination of the water supply is the least effective method for disseminating a biological weapon in these countries. Contamination of food supplies would most likely be used in a terrorist attack because it is difficult to contaminate enough food to gain a military advantage. Release of infected vectors is not particularly efficient for either military or terrorist purposes and entails a high probability of affecting those producing the weapons or living nearby.

By far the most effective mode for applying biological weapons is an aerosol cloud. Such a cloud is made up of microscopic particles and is therefore invisible. It can be produced in several ways, most of which involve either an explosion (some type of bomb) or spraying (usually involving a special nozzle on a spray tank). The effectiveness of the cloud is determined by numerous factors, such as the amount of agent that survives the explosion or spraying and the wind and weather conditions outdoors or air flow and ventilation indoors.

The primary result of an effective aerosol cloud is simultaneous infections among all those who were exposed to a sufficiently dense portion of the cloud. If the agent is contagious, the disease will then spread. In addition, agents that can survive for a long time in the environment will eventually settle, contaminating the ground, buildings, water and food sources, and so on. In some cases these sediments can form another dangerous aerosol cloud if they are disturbed.

Many countries have produced biological weapons for military use. The United States had a biological weapons program until 1969. Japan produced and deployed biological weapons during World War II. But by far the biggest and most sophisticated biological weapons program was that of the Soviet Union. Although the Soviet Union was a party to the 1972 Biological and Toxin Weapons Convention, it developed and produced biological weapons in huge quantities through at least the early 1990s. The size and scope of this program were enormous. For example, in the late 1980s and early 1990s, over 60,000 people were employed by the agencies responsible for the research, development, and production of biological weapons. Hundreds of tons of anthrax weapon formulation were stockpiled, along with dozens of tons of smallpox and plague. The total production capacity of all of the facilities involved was many hundreds of tons of various agents annually.

The Soviet Union's biological weapons program was established in the late 1920s. Before World War II, research was conducted on a wide variety of agents. But by the beginning of the war, the Soviet Union was able to manufacture weapons using the agents for tularemia, epidemic typhus, and Q fever and was also working on techniques for producing weapons using the agents for smallpox, plague, and anthrax. Analysis of a tularemia outbreak among German troops in southern Russia in 1942 indicated that this incident was very likely the result of the USSR's use of biological weapons. There was also a suspicious outbreak of Q fever in 1943 among German troops vacationing in the Crimea.

After the war, the Soviet program continued to expand and develop. While the prewar list of weaponized agents included tularemia, epidemic typhus, and Q fever, the postwar list was expanded to include smallpox, plague, anthrax, Venezuelan equine encephalomyelitis, glanders, brucellosis, and Marburg infection. Numerous other agents were studied for possible use as biological weapons, including Ebola, Junin virus (Argentinian hemorrhagic fever), Machupo virus (Bolivian hemorrhagic fever), yellow fever, Lassa fever, Japanese encephalitis, and Russian spring-summer encephalitis. Techniques and equipment were developed and refined for more efficient cultivation and concentration of the agents. Methods for producing dry weapons formulations for a number of agents also were developed. In addition to weapons to affect humans, a number of weapons to affect crops and livestock were developed.

During this postwar period, which lasted until the signing of the 1972 Biological and Toxin Weapons Convention, the Soviet Union also formulated its doctrine regarding the production and use of biological weapons. In the Soviets' definition, "strategic" weapons were those to be used on the deepest targets, that is, the United States and other distant countries; "operational" weapons were those intended for use on medium-range targets, nearer than the strategic targets but well behind the battlefront; and "tactical" weapons were those to be used at the battlefront. Biological weapons were excluded from use as "tactical" weapons and were divided into "strategic" and "operational" types. "Strategic" biological agents were mostly lethal, such as smallpox, anthrax, and plague; "operational" agents were mostly incapacitating, such as tularemia, glanders, and Venezuelan equine encephalomyelitis. For both types of weapons, use was envisioned on a massive scale to cause extensive disruption of vital civilian and military activity. The Soviets also established so-called mobilization capacities—facilities whose peacetime work was not biological weapons production but that could rapidly begin weapons production if war was imminent.

It is important to note that in the Soviets' view the best biological agents were those for which there was no prevention and no cure. For those agents for which vaccines or treatment existed—such as plague, which can be treated with antibiotics—antibiotic-resistant or immunosuppressive variants were to be developed. This is in sharp contrast to the philosophy of the U.S. program (terminated in 1969 by President Nixon's executive order), which stringently protected the safety of its biological weapons researchers by insisting that a vaccine or treatment be available for any agent studied.

After the Soviet Union became a party to the 1972 Biological and Toxin Weapons Convention, internal debate ensued about the fate of the existing biological weapons program. The end result was that the program was not dismantled but further intensified. During the period 1972 to 1992, the focus of the program was expanded. In addition to continuing previous types of work (developing improved manufacturing and testing techniques and equipment, developing improved delivery means for existing weapons, and exploring other possible agents as weapons), new emphasis was placed on the following:

- conducting molecular biology and genetic engineering research in order to develop antibiotic-resistant and immunosuppressive strains and to create genetically combined strains of two or more viruses;
- studying peptides with psychogenic or neurogenic effects as possible weapons;
- transforming nonpathogenic microorganisms and commensals into pathogenic microorganisms; and
- testing all of the facilities considered part of the "mobilization capacity" to verify their readiness.

As the Soviet Union weakened during the late 1980s and early 1990s, and as more and more details were revealed to the West regarding the Soviet biological weapons program, the West put increasing pressure on the Soviets. In 1991 a series of trilateral visits of biological facilities was conducted by the United States, Great Britain, and the Soviet Union. The Soviet biological weapons program still existed when these visits took place; the Soviets covered up the evidence as best they could.

After the collapse of the Soviet Union in early 1992, Russian President Boris Yeltsin signed a decree banning all biological weapons-related activity. Considerable downsizing in this area did indeed occur and included destruction of biological weapons stockpiles. However, there still remains doubt that Russia has completely dismantled the old Soviet program. There are some reasons to be concerned that biological weapons research and development are continuing in Russia today.

Russians have steadfastly refused to open their military biological weapons facilities to international inspection. The Russian biological weapons facilities that have received visitors have been those managed by the civilian arm of the Soviet/Russian biological weapons program, Biopreparat. The facilities of the Ministry of Defense, most notably those at Sergiyev Posad (formerly Zagorsk), Kirov, Yekaterinburg, and Strizhi, have never been visited.

Of course, Russia is not the only biological weapons threat the United States faces. A number of other states are known or suspected to possess biological weapons. A more immediate threat, though, is posed by potential terrorist use of biological weapons. The interest of terrorist groups in biological weapons is no surprise. Biological weapons have a number of very attractive features for terrorist uses. Their killing power can approach that of nuclear weapons. They are relatively inexpensive to make. A small-scale biological weapons attack using a common disease organism, such as tularemia or one of viral encephalitis, can be masked as a natural outbreak. The effects of a biological weapons attack are not apparent for several days, allowing the perpetrator time to vanish. The raw material—disease-producing strains of microorganisms—is fairly easy to obtain. And the techniques and equipment that are used in ordinary biotechnology research and production can be used for biological weapons.

Terrorists interested in biological weapons are on the level of state-sponsored terrorist organizations such as that of Osama bin Laden, on the level of large independent organizations such as Aum Shinrikyo, or on the level of individuals acting alone or in concert with small radical organizations. Although these groups will produce biological weapons with varying levels of sophistication, they all can potentially cause great damage. While the most obvious damages are illness and death, other potential results include panic; direct economic losses due to the costs of medical care, decontamination and other forms of cleanup, crowd control, and collateral agricultural damages such as animal deaths; and indirect economic losses caused by a drop in tourism and/or bans on farm exports from the target area.

Furthermore, there is no doubt that we will see future uses of biological weapons by terrorist groups, as there have been several attempts already. One incident, in 1984, involved members of the Rajneeshee cult contaminating restaurant salad bars in Oregon with salmonella, sickening 751 people. Another involved the Aum Shinrikyo cult. Although best known for its chemical attack in the Japanese subway system in 1995, the cult also attempted to release anthrax from the rooftop of a Tokyo building in 1993. No casualties resulted, but had the cult better understood cultivation of anthrax spores and urban air flow dynamics, the results might have been quite different.

Obviously, as illustrated by the difficulties Aum Shinrikyo experienced in mounting a biological weapons attack, it is not true that anyone who can brew beer can make a batch of biological weapons. Although someone with a strong background in microbiology could certainly produce a crude biological weapon to affect a small number of people and create panic, the production of sophisticated biological weapons requires sophisticated knowledge. For terrorist groups the most likely source of such knowledge would be state-sponsored biological weapons programs, which have the financial and scientific wherewithal to perfect production and deployment techniques. Because the former Soviet Union and Russia had the most sophisticated and powerful biological weapons program on earth, the former Soviet states present a particular proliferation threat. The tremendous knowledge amassed by former Soviet scientists would be extremely useful to both military and terrorist organizations.

When most people think of proliferation, they imagine weapons export. In the case of biological weapons, they picture international smuggling either of ready-made weapons material or at least of cultures of pathogenic microorganisms. However, this area of proliferation is of the least concern. Even without such assistance, a determined organization could obtain virulent strains of microorganisms from their natural reservoirs (such as soil or animals), from culture libraries that provide such organisms for research purposes, or by stealing cultures from legitimate laboratories. To the best of our knowledge, the former Soviet Union and Russia have not exported actual strains of microorganisms.

There are other types of biological weapons proliferation that are of greater concern. The first involves experienced scientists traveling or moving abroad. For example, there have been unconfirmed reports that scientists from the Kirov facility visited North Korea in the early 1990s. In addition, numerous scientists who used to work for the Soviet biological weapons program are now living abroad. Many of these scientists live in the West, but others have gone to Iran and other countries where their expertise can be put to nefarious use in state-run biological weapons programs. A second type of proliferation involves scientists from other countries being brought to a proliferating country for training in biotechnology, microbiology, and genetic engineering techniques. For years Moscow State University provided such training to scientists from dozens of countries, including Cuba, North Korea, eastern bloc nations, Iran, Iraq, Syria, and Libya.

A third form of proliferation involves private companies selling scientific expertise. For instance, a flier from a company advertises recombinant *Francisella tularensis* bacteria with altered virulence genes. Ostensibly, these organisms are being offered for vaccine production; the flier also notes that they can be used as genetic recipients and to create recombinant microorganisms of biologically active agents. The authors of the flier also express willingness to form cooperative ventures to which they will contribute their genetic engineering knowledge. The director of this company used to work for the USSR's biological weapons program.

A fourth type of proliferation occurs when a proliferating country sells equipment that can be used in biological weapons production. Such equipment is generally termed "dual-use," as it can be used for legitimate biotechnology production and for biological weapons production. An example of such proliferation was the planned sale by Russia of large fermenters to Iraq after the Persian Gulf War. Fortunately, the sale was not completed. We have no doubt that these fermenters were destined for use in biological weapons production. First of all, Iraq has used the guise of single-cell protein production as a cover for biological weapons facilities in the past. Second, the particular fermenter size involved in this proposed sale would not be suitable for efficient single-cell protein production. In fact, the resultant product would be prohibitively expensive. Similarly, in 1990, Biopreparat was negotiating the sale of dual-use equipment to Cuba as well.

The fifth kind of proliferation consists of published scientific literature. Just by reading scientific literature published in Russia in the past few years, a biological weapons developer could learn techniques to genetically engineer vaccinia virus and then transfer the results to smallpox; to create antibiotic-resistant strains of anthrax, plague, and glanders; and to mass produce the Marburg and Machupo viruses. Billions of dollars that the Soviet Union and Russia put into biotechnology research are available to anyone for the cost of a translator.

Given the current economic situation in the states of the former Soviet Union, the incentive to sell equipment and knowledge suitable for biological weapons production without regard to their eventual use is great for both the government and individual scientists and businessmen. The Russian government has long been short of funds, and its biotechnology arena has also been adversely affected. Many of its scientists are unemployed; those who are employed are paid poorly or not at all. Some have been forced to turn to other lines of work, such as street vending. It is important for the international community to ensure that these scientists have legitimate, decent-paying work to do in their fields.

The proliferation issue is particularly complex for biological weapons. In many cases the same equipment and knowledge that can be used to produce biological weapons can also be used to produce legitimate biotechnological products such as vaccines and antibiotics. Thus, we cannot outright forbid the export of most of the relevant knowledge and equipment as we can with nuclear weapons. Even if we did, such regulations would be practically impossible to enforce.

We believe that the United States should strive for transparency in the conduct of dual-use research and in the trade of cultures of pathogenic microorganisms and sophisticated biotechnology equipment. Clear international standards should regulate such trade. Such regulations would entail bans on certain activities, such as the sale of pathogenic microorganisms to individuals not associated with legitimate research institutions—not with the assumption that the ban would be enforceable but to clearly delineate acceptable conduct. The main focus of the regulations, though, would be reporting requirements for the sale or transfer of potentially dangerous cultures, genetic material, or equipment. An international organization would maintain the records of such transfers. While export controls, international treaties and inspection protocols, protective suits, and vaccines all play a role in the defense against biological weapons, none of these can eliminate the threat entirely.

Biological weapons are in essence a medical problem and thus require a medical solution. The ultimate goal of biodefense is to prevent suffering and loss of life. If biological weapons have minimal impact on the well-being of their targets, they are ineffective and thus cease to be a threat. Therefore, we must concentrate on developing appropriate medical defenses. There are three main types of medical defense against biological weapons: pretreatment (administered before exposure), urgent prophylaxis (administered after exposure but before symptoms arise), and chemotherapy (administered after the onset of illness). Pretreatment consists largely of vaccines but also includes certain drugs that can be administered before exposure to prevent disease. Use of pretreatment measures in biodefense will be effective only when all of the following conditions are met:

- The target population is known and limited—that is, military troops within range of an enemy's arsenal—since it is not realistic to vaccinate or provide drugs to everyone in the country.
- It is known precisely what biological agents are in the enemy's arsenal or the number of possible agents has been narrowed down to a few, since it is impossible to vaccinate or provide drugs against dozens of agents simultaneously.
- Pretreatment for the agent has already been developed. Note that for many biological agents, among them glanders, melioidosis, Marburg virus, Ebola virus, and Lassa fever, no vaccine exists; for most viral agents no pretreatment exists.
- The biological agents used are not genetically altered strains that are vaccine or drug resistant.

Clearly, pretreatment techniques are of very limited use. Therefore, we cannot rely exclusively or even primarily on pretreatment for medical biodefense. We must also ensure that the means for urgent prophylaxis and treatment of these diseases are available as well. Of the existing drugs that could be useful in urgent prophylaxis and treatment, many are not available in sufficient quantities; some are no longer manufactured. In addition, for many of the agents that can be used as biological weapons, no drug treatment protocols exist. The United States must greatly increase its efforts to develop new treatments and urgent prophylaxis techniques. This should include new approaches, such as preparations that can protect against and treat a wide variety of pathogens.

These efforts, as well as the funds spent on research and development, will pay for themselves many times over. In addition to contributing to preparedness for a biological attack, they will provide a much-needed push in the treatment of infectious diseases that occur under natural conditions. Infectious diseases remain one of the leading causes of death in the world, resulting in tremendous losses in terms of both money and human lives every year. Furthermore, such medical research would also contribute to the treatment of noninfectious diseases, such as autoimmune disorders and cancer.

The twenty-first century is anticipated to be the century of biotechnology and information technologies. This is a potent mix for future biological weapons development. The rapid advances anticipated in microbiology, molecular biology, and genetic engineering will improve our lives—but they are all dual-use technologies that can also be used in biological weapons development. Our improved knowledge of medicine and the functioning of the human body will enable us to improve human health and quality of life—but can be used to develop more sophisticated biological weapons. The explosive growth of information technology means that anyone with a computer has instantaneous access to tremendous amounts of information—including techniques that can be used in biological weapons development.

We cannot, and should not, halt the progress of science and technology, but we must bear in mind that it is a double-edged sword. To protect ourselves from the threat of biological weapons, we must increase our awareness and understanding of the threat, strengthen current international agreements, increase transparency, and most importantly, develop new medical means to render such weapons useless.

16

National Innovation to Combat Catastrophic Terrorism

Ashton Carter

What is the essential nature of international security in the post-Cold War world? And what is the American security strategy for this new period in time? The fact that we have no other term for this era than post-Cold War tells us that we know from whence we came but not where we are or where we are going. This uncertainty was acceptable in the early post-Cold War years when the task at hand was to manage the adjustment in a safe way. But a decade later we no longer have an excuse for not developing a better conception of a national security strategy.

Although there have been regional crises that have required military intervention by the United States—such as in Kosovo, Bosnia, Somalia, Rwanda, and Haiti—these situations do not directly affect U.S. vital interests. Moreover, the U.S. Department of Defense's (DOD) budget is largely defined by the need to be able to conduct two near-simultaneous major theater wars. This is the basis on which our force structure is sized and on which the preponderance of the defense budget is measured. However, although these important contingencies require readiness and would directly affect vital U.S. interests, they do not threaten the survival of the country, our way of life, or our position in the world. Problems of the highest order, those that make up the so-called A list, are those that do.

Today the A list is empty, which is a happy fact but one that could change in the not too distant future. The absence of immediate A-list threats creates a whole other dimension to national and international security that does not take the form of a traditional military threat and that therefore often goes unnoticed. Over time we have tended to conceive of our national security strategy as one that is based on perceived threats; that is, once identified, threats are properly addressed. But the threats posed by prospective items on the A list are not well defined and require a more preventive approach toward national security strategy.

In 1999, William Perry and I wrote a book entitled *Preventive Defense: A New Security Strategy for America* (The Brookings Institution, Washington, D.C.) in which we intended to make the analogy to preventive medicine. That is, while curative medicine takes care of people once they are sick, preventive medicine keeps them from getting sick in the first place. And while curative defense may be appropriate for situations that are not on the A list, preventive defense is what is needed for situations that are on the list. As riveting as the crises in such places as Kosovo are, we were concerned that as a nation the United States is losing sight of preventive defense.

THE A LIST

Among the A-list problems that William Perry and I identified as requiring preventive measures is the state of affairs in Russia. Of particular concern is whether Russia develops a Weimar Russian syndrome, in which it believes it is being treated in a disparaging way, much as Germany did after World War I. The danger of such a response would be the resulting embitterment and isolation. Another area of concern is the emergence of a more prominent China and its subsequent geopolitical effects.

A third issue central to the theme of this colloquium is what we call "catastrophic terrorism" or "grand terrorism," of which bioterrorism is probably the most menacing variety. By catastrophic terrorism we mean an attack outside the context of traditional warfare but with war-scale consequences. Such an attack might take the form of biological, chemical, nuclear, or cyber warfare unleashed on the critical infrastructures that allow our complex and interdependent societies to function. Catastrophic terrorism is orders of magnitude more serious than hostage taking, bombs in the marketplace, or airplane hijackings. In other words, we are not talking about the destruction of a building in Oklahoma City but of Oklahoma City itself, not just disruption and some damage to the World Trade Center towers in Manhattan but the total destruction of both buildings and the surrounding area.

Just because these attacks have never happened does not mean that they cannot or will not. With a little protection, though, we can make the possibility even smaller. Unfortunately, this destructive potential is falling into the hands of smaller and smaller groups of people, an inevitable consequence of the passage of time and the rising tide of technology. We are also becoming increasingly vulnerable as societies. International networks of money laundering, crime, and drug peddling abound, providing an infrastructure for transnational terrorism. In addition, our society includes groups with messianic or vengeful motivations. These kinds of groups have nothing to lose and are the ones most likely to resort to grand terrorism. Conversely, groups with a political agenda that hope to become legitimate political forces are not likely to resort to grand terrorism.

Because no country or group in the world can take on the United States in a frontal way militarily, some groups might believe they have no option but to attack us in an "asymmetrical" manner through terrorism. Such an incident would abruptly undermine Americans' sense of security on their own soil. In America's experience, security emergencies occur elsewhere. The last time we felt vulnerable in our own country was in 1949, when Joseph Stalin exploded an atomic weapon and it became clear that cataclysmic damage could be inflicted on the United States. That began a 50-year process of trying to understand and prepare for an emergency of such magnitude while simultaneously preserving the American way of life.

The second aspect of such an attack is its likely transnationality. Terrorism is not a problem that will respect borders. It will be a threat from within as well as from without, and it raises the possibility that the circles of violence will quickly widen. Imitators will try to replicate an early success, and in our search for those responsible we are likely to set aside treasured civil rights. Moreover, the likelihood that we will fail to prevent this incident and respond to it properly will discredit the authorities in the eyes of the public. We need only be reminded of German terrorists in the 1970s and 1980s, whose objective was to disgrace the public order, to demonstrate the German government's inability to protect its own population, and/or to provoke the German government to take repressive steps that would in turn undermine its authority.

PREVENTING CATASTROPHIC TERRORISM

A number of national security law enforcement experts, intelligence officers, and scientists have begun to address the issues of building an architecture for response to catastrophic terrorism in this country. It will take decades to adjust to this new reality and to comprehend the dimensions of the problem. We must consider the entire spectrum of prevention, ranging from broad to focused surveillance (which involves narrowing down on a group or installation), to tactical and strategic warning, to psychological prevention strategies, to deterrence.

We also need to develop various protections, including vaccines, masks, interdiction of terrorism, and detection of terrorist acts as they are being prepared or executed, and we need to plan to clean up after such an attack and to develop a forensic plan for determining the culprit. An important dimension of prevention and response is the multidisciplinary and multiagency nature of the need. How the event is labeled—as an attack, as a crime, or as a disaster—will determine who takes the lead. Like many new significant challenges to governance, this problem cuts across the lines we have drawn for conducting public business and argues for the establishment of a central coordinating office—a "czar." But first the capability and infrastructure must be developed.

A strategy that addresses catastrophic terrorism includes many components. The first and most sensitive one is intelligence, which if mishandled can cut right to the heart of our civil liberties. Transnational terrorism falls into the chasm that is deeply carved in our government's topography between the national security and law enforcement communities. One deals with threats to our security, the other with crime. For two centuries we have tolerated this sharp divide because security threats have always come from the outside. The rules of engagement in dealing with terrorists change on foreign soil. For example, the national security community uses proactive intelligence—using all means of surveillance—to identify threats and take vigorous preemptive action. In contrast, our criminal justice system has limited powers of surveillance and is supposed to target Americans only after just cause is established or a crime is committed.

Thus, attempts to engage the Federal Bureau of Investigation and the Central Intelligence Agency in cooperative efforts aimed at counterterrorism result in a clash of cultures. The challenge is to find some means by which information collected by the intelligence community, which does not want that information revealed in court, can be provided to law enforcement, which requires that the information that is used as a basis for prosecution be divulged.

Another sensitive issue concerns the role of DOD. Is this national defense? Is this a theater of war? Should DOD be involved? Many find DOD involvement threatening because they do not think of American soil as an appropriate place for military activity. DOD recently designated a commander whose responsibility is to manage its contribution to a domestic terrorism situation.

A third issue is the importance of planning well in advance of an actual incident. Now is the time to try to develop balanced approaches to these difficult problems; the day after an attack is too late. Indeed, we should fear an ill-conceived and panicky response to a catastrophic terrorism incident as much as we fear the incident itself. In addition, we must consider making biological warfare a universal crime. The Biological Weapons Convention makes it a crime under international law for states to conduct biological warfare. We also must make it an individual crime in all states to engage in biological warfare. Piracy and airline hijacking provide two examples of precedent in this area.

Finally, we must educate our legislators, many of whom are locked into the paradigm of conventional weaponry, about the threat of biological warfare. This will be a critical aspect of prevention because without legislative support the necessary resources will not be available.

BIBLIOGRAPHY

Carter, A. B., J. M. Deutch, and P. D. Zelikow. 1998. Catastrophic terrorism: A new national policy. Foreign Affairs, 77(6):80-94.

Carter, A. B., J. M. Deutch, and P. D. Zelikow. 1998. Catastrophic terrorism: Elements of a national policy. Preventive Defense Project Publications, 1(6).

Carter, A. B., and W. J. Perry. 1999. A false alarm: This time. In Preventive Defense: A New Security Strategy for America. Washington, D.C.: The Brookings Institution.

17

Flow Cytometry Analysis Techniques for High-Throughput Biodefense Research

James H. Jett, Hong Cai, Robert C. Habbersett, Richard A. Keller, Erica J. Larson, Babetta L. Marrone, John P. Nolan, Xuedong Song, Basil Swanson, and Paul S. White

INTRODUCTION

Contemporary flow cytometry has its roots in developments of rapid cellular analyses that began in the 1960s. Initial measurements focused on the analysis of hematopoietic and other mammalian cell types to determine size distributions, DNA content distributions, and surface antigen expression distributions (Melamed et al., 1990). With the advent of monoclonal antibodies, studies of immune systems by flow cytometry exploded, and important fundamental aspects are still being discovered. Over the years, the application of flow cytometry has moved from the analysis of whole cells and their constituents to that of subcellular components. Through improvements in detection sensitivity, it is now possible to analyze individual molecules (Ambrose et al., 1999). Thus, the name "cytometer," or cell analyzer, is not appropriate today. A more descriptive name for the technology is "fluorescent particle analysis in a flowing sample stream." Several applications of the basic technology to high-throughput research are emerging in the biodefense arena.

The key to understanding the capabilities of flow cytometric measurements is contained in the method of sample presentation. In a generic sense, objects in suspension to be analyzed are introduced into the measurement region after hydrodynamic focusing by a surrounding sheath fluid. This is depicted schematically in Figure 17.1. In addition to confining the sample stream to a very narrow diameter, typically 10 microns, the sample velocity is increased. The fully focused sample stream passes through one or more focused laser beams. In addition to scattering light,

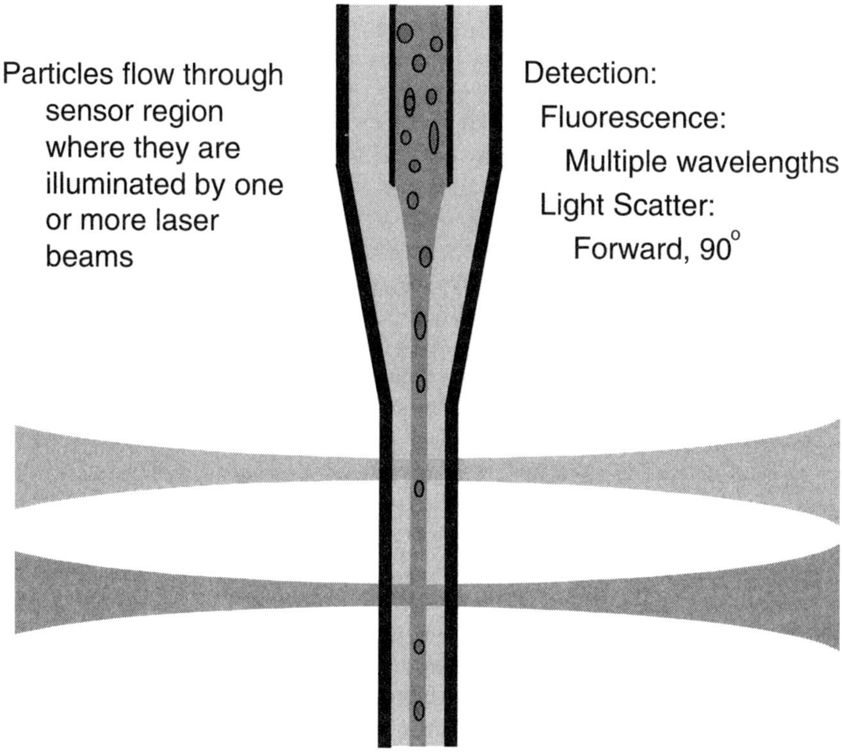

FIGURE 17.1 A schematic drawing of a typical flow cytometer illustrating the single-particle sample inlet tube at the top and the hydrodynamic focusing of the sample stream as it is transported through one or more laser beams. Fluorescence is collected at right angles to the laser beam sample stream axes, filtered optically, and focused onto detectors.

fluorescent dyes bound to specific components of the objects being analyzed are excited and emit fluorescence photons. The laser beam transit time is roughly dependent on the measurement sensitivity required. For cellular analyses, the transit time is typically 10 microseconds. This translates into analysis rates ranging up to tens of thousands of particles per second. For single-molecule detection and DNA fragment size analysis, the transit times are measured in milliseconds, which limits the analysis rates to roughly 100 molecules per second.

The fluorescence photons emitted as the particles pass through the laser beam are collected, filtered spectrally, and directed to one or more detectors. Detector signals are processed to derive up to three types of information: transit time through the laser beam, maximum fluorescence emission, and total fluorescence emission. These pieces of information from each detector are correlated with other measurements, such as those listed below, for the particle and recorded in the computer. Subsequent offline analysis of the multidimensional data collected is often required to extract the biological information of interest.

For the higher-velocity cell analysis systems, it is possible to physically separate (sort) the particles by forcing the stream exiting the flow cell to form droplets at an induced frequency. The optimal frequency is roughly inversely proportional to the exit orifice diameter, ranging from 5 kHz for a 200-micron-diameter orifice to 70 kHz for a 50-micron-orifice. Measurements made as the particles transit the laser beam(s) are used to define which particles are to be separated from the parent population. When a particle whose measurements satisfy the sort criteria is contained in the terminal droplet that is ready to break off from the solid stream above, a charge is applied to the stream. Once the droplet has broken off, the charge is removed from the stream, but the droplet remains charged as it falls between deflection plates that are held at a constant high voltage. The deflected droplets are collected in a vessel, while the particles not of interest are not deflected.

In addition to measuring the total or maximum fluorescence emitted in multiwavelength regions, other properties of the fluorescence emissions have been measured, including polarization state and lifetime. Scattered-light measurements, when properly analyzed, yield information about a particle's size and surface characteristics.

Two examples of flow cytometric measurements that have application in the biodefense arena are DNA fragment size analysis and microsphere-based assays for multiplexed minisequencing and toxin detection. Currently, the first two of these technologies are being developed to identify bacteria, bacterial strains, and bacterial pathogenicity.

DNA FRAGMENT SIZING FLOW CYTOMETRY

DNA fragment size distribution analysis is a ubiquitous measurement in molecular biology. Applications exist in disease diagnostics, large insert DNA clone analysis, bacteria strain and species identification, polymerase chain reaction (PCR) product analysis, and forensics. Based on our experience in individual fluorescent analysis (Ambrose et al., 1999), we have developed a new technique for sizing DNA fragments based on flow cytometry (Goodwin et al., 1993; Petty et al., 1995; Habbersett et al., 2000). For large DNA fragments—fragments greater than 30 kilobase pairs—our approach is more sensitive (sample size of femtograms versus micrograms), faster (analysis time of 5 minutes versus tens of hours), and more accurate (size uncertainty of 2 percent versus 10 percent) than pulsed-field gel electrophoresis (PFGE), commonly used for these analyses.

State-of-the-art techniques for DNA fragment size analysis use some form of electrophoresis. For samples containing DNA fragments greater than 30 kilobase pairs in length, PFGE methods are used that require up to 24 hours or more to achieve the desired separations of hundreds of nanograms of DNA. Furthermore, all forms of electrophoretic separation are highly nonlinear, which affects the accuracy of fragment size measurements.

PFGE is often used in hospital settings and in state public health laboratories to determine the presence of an outbreak or to make a diagnosis of a bacterial infection. Currently, sample preparation methods are carried out in a plug of the gel matrix in which the bacterial sample is embedded and processed. Due to long diffusion times of reagents into and out of the plug, the process can require up to 7 days to complete before the PFGE analysis. Recent developments at Los Alamos and elsewhere (Chang and Chui, 1998) have reduced the sample preparation time to 8 hours or less.

In general, a bacterial sample is embedded in an agarose matrix and reagents are diffused into and out of the agarose that break down the bacterial cell wall, digest the proteins, and cut up the bacterial genome with restriction enzymes. The restriction endonuclease severs the bacterial DNA at specific sequence sites. Typically, for bacterial species and strain identification, the restriction sites are eight bases long, which results in an average fragment length of 65 kilobase pairs. Eight-base instead of more frequent cutters are used to keep the number of fragments produced per genome manageable. Analysis of the distribution of fragment sizes by PFGE or flow cytometry produces a "fingerprint" pattern that is specific to the bacterial species and strain being analyzed. Often, the restriction digest fingerprint can uniquely identify the organism.

For analysis of bacterial genome restriction digests by flow cytometry, the DNA sample, which is prepared in the same manner as for PFGE analysis, is stained with a fluorescent intercalating dye that binds stoichiometrically to the DNA such that the amount of dye incorporated is directly proportional to the fragment size (number of base pairs [bp]). The 1,000-fold increase in the fluorescence quantum yield of the intercalating dye upon binding to the DNA makes it unnecessary to remove unbound dye from the solution before analysis. The stained fragments are diluted to 10^{-12} to 10^{-14} molar and introduced into an ultrasensitive flow cytometer. Fragments pass individually through the laser-illuminated detection region of the flow cytometer, each fragment producing a burst of fluorescence photons as it transits the laser beam. Fluorescence bursts from individual fragments are quantified and recorded. The burst size is a measure of the fragment size. A histogram of the burst sizes is generated that displays the distribution of fragment sizes in the sample (i.e., a DNA fingerprint). The response linearity has been verified stepwise over the size range analyzed to date: 125 to 450,000 bp.

Histograms of the burst sizes resulting from the analysis of restriction digests of four *Staphylococcus aureus* strains are shown in Figure 17.2. Each histogram was compiled from less than 5 minutes of data (15,000 events) acquired from less than a femtogram of DNA. Plots of the burst sizes

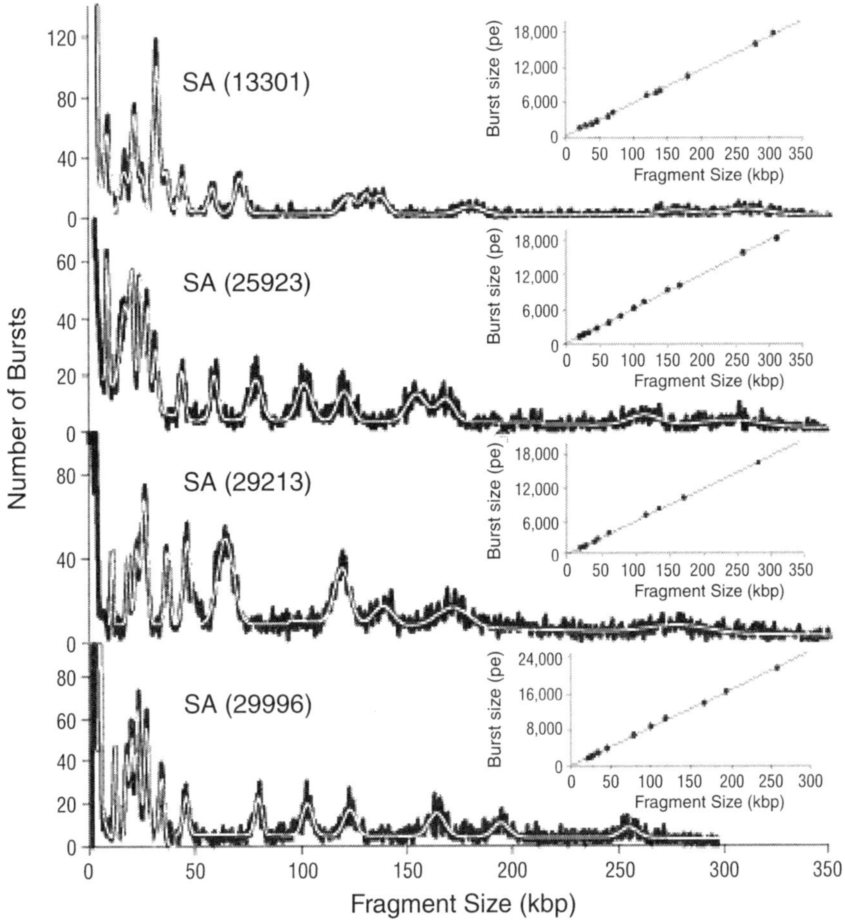

FIGURE 17.2 Examples of the restriction digest fingerprint patterns by four strains of *S. aureus*. The inserts show the relationship between the fragment sizes measured by flow cytometry and the sizes determined by PFGE.

versus the fragment lengths determined by PFGE analyses also are displayed. The plots are linear with correlation coefficients of $r^2 > 0.999$. Most of the deviation of the points from the fit line is attributed to the 10 percent uncertainty inherent in PFGE versus the 2 percent uncertainty of flow cytometric measurements. Each strain produced a distinct fingerprint. Similar results have been obtained for several strains of *Escherichia coli* (Larson et al., 2000).

Application of this technology to bacterial species and strain identification is being pursed vigorously. Macrorestriction digests of bacterial genomes with infrequently cutting restriction endonucleases analyzed by PFGE provide one of the most definitive signatures for species and strain identification (Maslow et al., 1993a, 1993b). These analyses typically require micrograms of DNA, 8 hours for sample processing, followed by ~20 hours for PFGE separation and analysis of the results. The patterns observed, which vary with restriction enzyme used as well as with species and strain analyzed, form the basis for identifying the bacterial species and strain.

A description of species discrimination by flow cytometric DNA fragment sizing has been published (Kim et al., 1999). Figure 17.2 demonstrates an example of flow cytometric measurements of restriction digests of four strains of *S. aureus* digested with the restriction enzyme Sma I. In addition to sample preparation time being reduced to 6 hours followed by 5 minutes of analysis, the amount of DNA required is one-millionth the amount needed for PFGE analysis.

There are several specific biodefense applications of DNA fragment sizing technology. The technology is capable of recognizing and identifying both bacterial species and strains. This capability is useful in forensic applications, where attribution to a source of a biothreat agent used in an incident is desired, in public health laboratories for identifying and tracking outbreaks, and at special events to confirm the identity of biothreat agents by other means. For nonproliferation assays it will be possible to develop profiles of the effluents from production facilities and to confirm organism identification in treaty verification studies. Highly automated throughput sample preparation and analysis systems will be developed in the future to enable the processing and analysis of large numbers of samples.

FLOW CYTOMETRIC MICROSPHERE-BASED ASSAYS

A characteristic of a conventional flow cytometer is that, since the sample is confined to move in a narrow stream and the excitation laser is tightly focused, the volume that is observed by the detection optics can be small (pL – nL), and fluorescence emissions originating outside the probe volume are not "seen" by the detector. The operational result of this characteristic is that it is not necessary to separate particle-bound from free fluorescent molecules, which enables the performance of a wide variety of homogeneous assays.

Microspheres in the 3- to 10-micron diameter range are used to carry the molecular assemblies for a number of biomolecular assays that use a flow cytometer as a readout instrument. Applications of microsphere-based assays include measurement of enzyme activity (Frank et al., 1998), cholera toxin detection (Song et al., 1998), immunoassays (Saunders et al., 1985), and single nucleotide polymorphism (SNP) detection (Cai et al., 2000). SNP detection is an example of the application of minisequencing where the reagents are carried on a microsphere.

In Figure 17.3 the circle represents a microsphere to which is attached a large number (~10^5) of oligonucleotides that end one position short of the site of interest. DNA from the sample (beginning and ending with . . .) being analyzed is hybridized to the oligonucleotides attached to the microsphere. The "C" is added to the top oligo by enzymatic incorporation of a dideoxy nucleotide that is fluorescently labeled. This reaction is carried out in four separate tubes, each containing a different labeled nucleotide, along with three unlabeled dideoxy nucleotides. Thus, when each of the four samples is analyzed, only one sample will be measured as having fluorescence bound to the microsphere—the one with the correctly labeled nucleotide incorporated. As biothreat organisms are sequenced, this technique will be used to determine agent species and strains.

The microsphere-based cholera toxin assay uses a different principle, that of fluorescence resonant energy transfer. When cholera toxin binds to a cell, multiple receptor molecules bind to the toxin. In this assay the microspheres are coated with a lipid bilayer in which are embedded two types of cholera toxin receptor molecules. The receptor molecules are labeled with either a fluorescence donor or acceptor. In the absence of the toxin, the donor-acceptor-labeled molecules are far enough apart that, when the donor molecule is excited by an appropriate wavelength of laser light, fluorescence energy transfer does not occur, resulting in a fluorescence emission spectrum from the bead that is characteristic of the donor molecule. However, when toxin is present, donor acceptor pairs are

FIGURE 17.3 Schematic representation of the minisequencing process.

brought into close proximity (~5 nm), and the fluorescence emission spectrum becomes more characteristic of the acceptor molecule. By spectral selection the ratio of the donor to acceptor fluorescence can be measured and, once calibrated, provides a measure of the cholera toxin concentration (Song et al., 1998).

FUTURE DIRECTIONS

To this point, single assays that are bead based have been discussed. A new technology has been developed that enables multiplexing of assays. A bead set developed by the Luminex Corporation (Austin, Tex.; *www.luminexcorp.com*) consists of beads that are stained with two dyes in varying amounts. The bead types are distinguishable by measuring the amount of fluorescence in two wavelength regions. The assay results, such as the minisequencing assay, are reported in a third wavelength region. Thus, a flow cytometer capable of recording fluorescence emissions in three wavelength regions is able to separate the bead types and record the assay results reported by each type. Currently, bead sets containing 64 bead types are available, with larger sets planned for the future. In addition to distinguishing beads by the amount of labeling dyes contained in them, it is possible to envision adding size, and perhaps shape, to provide other dimensions for multiplexing.

For high-throughput flow cytometric assays, automated instrument operation including sample introduction will be necessary. A recent development (Edwards et al., 1999) in sample introduction has achieved a sample throughput rate of nine samples per minute with less than 4 percent overlap between samples. The approach used is basically to line samples up in the sample delivery tube of a cytometer separated by regions of fluid that contain no particles. A series of computer-controlled valves coupled to a syringe drive are used to deliver the samples continuously to the cytometer. This and other developments in the future will enable truly high-throughput sample preparation and analysis that will take advantage of the flow cytometer's ability to analyze large numbers of individual particles in a sample at rates of thousands per second.

ACKNOWLEDGMENTS

This work was supported by the U.S. Department of Energy (NN-20), the National Institutes of Health National Center for Research Resources (Grant RR-01315), and the Los Alamos National Laboratory Laboratory Directed Research and Development program.

REFERENCES

Ambrose, W. P., P. M. Goodwin, J. H. Jett, A. Van Orden, J. H. Werner, and R. A. Keller. 1999. Single molecule fluorescence spectroscopy at ambient temperature. Chemical Reviews, 99:2929-2956.

Cai, H., P. S. White, D. C. Torney, A. Deshpande, Z. Wang, B. Marrone, and J. P. Nolan. 2000. Flow cytometry-based minisequencing: A new platform for high-throughput single-nucleotide polymorphism scoring. Genomics, 66:135-143.

Chang, N., and L. Chui. 1998. A standardized protocol for the rapid preparation of bacterial DNA for pulsed-field gel electrophoresis. Diagnostic Microbiology and Infectious Disease, 31:275-279.

Edwards, B. S., F. Kuckuck, and L. A. Sklar. 1999. Plug flow cytometry: An automated coupling device for rapid sequential flow cytometric sample analysis. Cytometry, 37:156-169.

Frank, Q., M. Somsouk, Y. Weng, L. Somsouk, J. P. Nolan, and B. Shen. 1998. Partial functional deficiency of E160D flap endonuclease-1 mutant in vitro and in vivo is due to defective cleavage of DNA substrates. Journal of Biological Chemistry, 273:33064-33072.

Goodwin, P. M., M. E. Johnson, J. C. Martin, W. P. Ambrose, B. L. Marrone, J. H. Jett, and R. A. Keller. 1993. Rapid sizing of individual fluorescently stained DNA fragments by flow cytometry. Nucleic Acids Research, 21:803-806.

Habbersett, R. C., J. H. Jett, and R. A. Keller. 2000. Single fragment detection by flow cytometry. Pp. 115-138 in Emerging Tools for Cell Analysis: Advances in Optical Measurement, J. P. Robinson and G. Durak, eds. New York: Wiley-Liss.

Kim, Y., J. H. Jett, E. J. Larson, J. R. Pentilla, B. L. Marone, and R. A. Keller. 1999. Bacterial fingerprinting by flow cytometry: Bacterial species discrimination. Cytometry, 36:324-332.

Larson, E., J. R. Penttila, H. Cai, J. H. Jett, S. Burde, and R. A. Keller. 2000. Rapid DNA fingerprinting of pathogens by flow cytometry. Cytometry, 41:203-208.

Maslow, J. N., M. E. Mulligan, and R. D. Arbeit. 1993a. Molecular epidemiology: Application of contemporary techniques to the typing of microorganisms. Clinical Infectious Diseases, 17:153-164.

Maslow, J. N., A. M. Slutsky, and R. D. Arbeit. 1993b. Application of pulsed-field gel electrophoresis to molecular epidemiology. In Diagnostic Molecular Microbiology: Principles and Applications, D. H. Persing, T. F. Smith, F. C. Tenover, and T. J. White, eds. Washington, D.C.: American Society for Microbiology.

Melamed, M. R., T. Lindmo, and M. L. Mendelsohn, eds. 1990. Flow Cytometry and Sorting. New York: Wiley-Liss.

Petty, J. T., M. E. Johnson, P. M. Goodwin, J. C. Martin, J. H. Jett, and R. A. Keller. 1995. Characterization of DNA size determination of small fragments by flow cytometry. Analytical Chemistry, 67:1755-1761.

Saunders, G. C., J. H. Jett, and J. C. Martin. 1985. Amplified flow cytometric separation free fluorescence immunoassay. Clinical Chemistry, 31:2020-2023.

Song, X., J. P. Nolan, and B. Swanson. 1998. An optical biosensor based on fluorescence self-quenching and energy transfer: Ultrasensitive and specific detection of protein toxins. Journal of the American Chemical Society, 120:11514-11515.

18

Forensic Perspective on Bioterrorism and the Proliferation of Bioweapons

Randall S. Murch

INTRODUCTION

The threat and dangers presented by the potential use of biological weapons as instruments of terrorism and warfare against the United States are real. Important recent discussions confirm the conclusion that nation states and other actors have placed large investments in the development and possession of biological organisms or their byproducts for nefarious purposes (Alibek and Handelman, 1999; Anderson, 1998; Commission, 1999; Falkenrath et al., 1998; Kaplan, 1998; MacKenzie, 1998; Rimmington, 1999). The possible public health implications have been described at length (Bardi, 1999, Bartlett, 1999; Franz, 1999; Kortepeter and Parker, 1999; Hamburg, 1999; Henderson, 1999; Hughes, 1999; Inglesby, 1999, O'Toole, 1999; Pavlin, 1999; Siegrist, 1999). Although some analysts debate the likelihood and extent to which bioweapons would be used against the United States (Stern, 1999; Tucker, 1999), the motive, technology, source materials, and knowledge—at least for what might be termed crude attacks—are widely available and accessible.

Over one dozen nations are suspected or known to have biological weapons development programs. Many are, or have been, openly hostile to the United States. The former Soviet Union had an extraordinarily well-developed program and since its "dismantlement" in 1992 has had very little or no control over the loss of seed cultures or experts to other countries or to the lure of entrepreneurism (Alibek and Handelman, 1999). Further, the recent rash of anthrax and ricin incidents and hoaxes in the United States being investigated by the Federal Bureau of Investigation

(FBI) and other authorities should remind us to take these dangers seriously. Thus, the list of possible actors expands to right-wing militia and white supremacist groups, those loosely affiliated with foreign terrorist groups, religious and millennialist groups, and lone perpetrators. We might do well to reflect often on former Senator Sam Nunn's well-known 1996 quote about bioterrorism: "It is not a matter of *if*, but a matter of *when*." An essential part of our defense against the illicit possession or use of biological weapons should be a national bioforensics network. Similar networks could be established for other weapons of mass destruction (WMD).

THE NEED FOR A NATIONAL BIOFORENSICS NETWORK

For just a few moments, step into the shoes of those responsible for protecting against or responding to these threats. Consider the topology they must plan for and prepare against. One dimension might include the array of hazardous materials that could be employed, accessibility to source or seed materials, and availability of relevant information, while another might include the possible delivery means and scenarios. In still another the categories of targets that exist, whether viewed as locations, environments, events, populations or infrastructures, might be found, and yet another might include level of sophistication, operational capabilities, psychology, motivations, categories, and numbers of adversaries. Still another dimension is that of time and distance. The threat matrix soon becomes daunting, if not overwhelming.

Over the past four years great concerns have been raised and expressed about gaps and shortfalls in U.S. preparedness and response to and recovery from the illicit use of biological agents. Most often these concerns arise in the context of bioterrorism conducted either at home or abroad. As a result, substantial funding has flowed forth in an attempt to address various aspects of these deficiencies, ranging from supplying personal protective equipment, instruments, and training to "first responders" (Pavlin, 1999), to research on advanced decontamination methods, to suggested new approaches for improving or boosting individual broad-spectrum immunity against disease agents (Alibek and Handelman, 1999), to stockpiling and improving vaccines and diagnostics against threat organisms (Anderson, 1998; Clarke, 1999; Hamburg, 1999; Pavlin, 1999). Clearly, full and robust defense and recovery from bioattacks must be a national priority; however, it should not be the only priority. There is another important chapter to this story that needs to come to life, namely a national bioforensics network.

This nation will not only expect its government to provide protection from and ensure complete and lasting recovery from a bioterrorism attack but will also demand the rapid and accurate identification of the cause, the means, and those involved. It will further expect that subsequent attacks will be prevented, that full and robust attribution is achieved, and that the perpetrators will be brought to justice, either through the U.S. courts or a political or military action. If we concentrate resources only on the medical consequences and disaster relief aspects of a bioattack, we will not be positioning ourselves properly against our adversaries or necessarily know who attacked us. Fully competent attribution supports investigation, resolution, and prosecution. The law enforcement and intelligence communities have significant responsibilities in this domain.

Over the past four years volumes of written and spoken words have been devoted to the consequence management aspects of bioterrorism and the proliferation of biological weapons. However, insufficient attention has been given to enhancing our ability to effectively predict, locate, disrupt, prevent, investigate, resolve, and deter such activities, either after the fact or, more importantly, before the fact. Obviously, several resource and policy implications must be considered, including one that is perhaps not so obvious. These considerations include:

- Increasing the number of trained intelligence and law enforcement personnel who collect, investigate, and analyze information as well as operate against WMD and those who possess them.
- Providing those individuals with better information management tools to make them more efficient and effective in finding and using relevant open- and closed-source information.
- Producing new technical capabilities via a national integrated research and development strategy to better detect, locate, identify, frustrate, disrupt, and defeat WMD production and use.
- Developing robust national policies for actions that would be taken against those who use biological weapons.

The not-so-obvious implication noted above is the critical role of attribution and forensics in response to and defense against WMD.

Given how a biological attack is likely to unfold—as an unobtrusive, unexpected release of an organism or toxin with effects detected days to weeks later—traditional law enforcement, intelligence, or even medical methods alone will very likely be insufficient. Forensic evidence will play a crucial role in investigating and solving the bioterrorism crime. Perhaps it will also play an important role in counterproliferation or even nonproliferation. For the purposes of this paper, forensics does not simply mean characterization of material to determine what it is; rather the more traditional definition is used, which also has the critical ingredient of relevance. This often requires that the origin of the sample be determined to a high degree of scientific certainty, which involves the use of highly discriminating analyses that compare samples from known and questioned sources.

THE FBI'S ROLE

Beginning in 1995, the FBI laboratory began articulating and addressing the need for integrating and advancing forensics capabilities for WMD investigation and resolution. Based on decades of experience with criminal and terrorism investigations and the investigation and prosecution of terrorist bombings, the FBI realized that an important void existed. No one had considered that forensics investigation would play a critical role in responding to WMD. Thus, in 1996 the FBI created the unique Hazardous Materials Response Unit (HMRU), which focuses on forensics investigation of such events (Ember, 1996). HMRU is an operational responder that quickly provides forensics leadership and integrates required expertise from the FBI laboratory and other agencies for on-scene commanders. Since its inception, HMRU has responded to dozens of suspected WMD (both criminal and terrorism) cases and is investigating many more.

HMRU and FBI laboratory management have also pursued aggressive outreach with a number of federal and professional scientific and medical organizations, which has resulted in merging forensics thinking, requirements, plans, and operations into theirs. The HMRU and its counterpart Bomb Data Center (advanced render safe) have been included in many interagency planning activities and exercises. Additionally, through joint efforts, various first-responder communities (law enforcement, fire/rescue, medicine, public health, hazardous materials) increasingly are becoming sensitive to requirements for the protection, preservation, collection, and analysis of physical evidence and will incorporate these into their practices and operations whenever possible. Evidence response teams in FBI field offices are being trained and equipped to respond to WMD crime scenes in order to provide more immediate response while the HMRU and its partners mobilize. HMRU has a small research budget that is focused on advancing field detection and laboratory analysis technologies and methods for biological and chemical agents.

REMAINING NEEDS

We are still a long way from having the needed forensics capabilities and resources. Research, development, and validation in WMD forensics must be expanded considerably. Increased resources need to be appropriated to accelerate the training, equipping, and exercising of WMD crime scene responders and forensics experts. Advanced forensics facilities must be constructed or provided to accommodate the analysis and storage of contaminated evidence. Logically, these should be in close proximity to or easily accessible by centers that provide related capabilities. Mobile scientific response should be optimized to permit the maximal amount and array of analyses to be performed in the field. An integrated laboratory architecture for WMD forensics should be established that builds on existing capabilities in key federal agencies.

Such traditional fields as latent fingerprint, trace evidence, computer forensics, DNA analysis, document examination, materials analysis, and forensics informatics routinely develop critical and timely leads and contribute evidence toward identifying perpetrators, scenes, instruments and methods of crimes. This would very likely also be true with crimes involving WMD and, specifically for the purpose of this paper, biological agents or toxins. A lack of resources has prevented the development or modification of technologies and methods to validly apply traditional forensics science to physical evidence that may be contaminated with a highly infectious or otherwise dangerous biological or chemical agent. It is the experience of the forensics community that considerable validation is required of novel forensics approaches *before* their use on probative evidence. Reversing this order means almost certain exclusion of key trial evidence.

Those of us involved in the scientific investigation of WMD bioevents also envision that medical microbiology, epidemiology, advanced microbial genomics, and bioinformatics must play a crucial role in attributing and prosecuting these crimes. We are not alone in this proposition (Fox, 1998; Snyder, 1999). The organism or toxin itself should be viewed not only as the causative agent of an unnatural disease outbreak but also as evidence in a crime. The forensics perspective is crucial. The forensics biologist (as well as investigators and attorneys or, alternatively, the FBI, the intelligence community, and the White House) would seek to answer the following questions:

- What is the material in question?
- What is its relationship or relevance to the incident being investigated?
- What is the source of the material?
- Can the source of the material in question be individualized or at least significantly narrowed?
- Can a link be established between the crime and the perpetrators? If so, how strong is that link?

The better the answers to these questions, in series, the greater the probative value of the evidence.

Our position would be much stronger if we were able to fully and robustly identify and characterize a bioweapon and if we were able to establish a link between those involved and the victims, crime scenes, and methods and tools of the crime. Identifying and uniquely determining the origin of a microorganism in a bioterrorism case would have the same high probative value that is achieved when, in a sexual assault case, semen from the accused is identified using evidence collected from the victim. With current technology, forensics scientists often achieve statistically robust attributions of biological evidence, for which the probability of a random match is less than 1 in 260 billion. By convention this is considered to be absolute identification. Exclusions of suspected sources, which are of equal or greater importance, are usually made very swiftly and protect the accused from false allegations, conviction, or punishment. Because of its power and impact, forensics DNA technology is widely considered to be the greatest scientific advancement in the U.S. criminal justice system in the last century. One could easily envision the impact of a similar capability for the type of biological evidence relevant to this discussion.

COLLABORATION AT THE FEDERAL LEVEL

Building on FBI laboratory leadership in forensics DNA analysis, we have begun to explore the potential of bioforensics with the U.S. Department of Energy, the U.S. Army Medical Institute of Infectious Disease, the Centers for Disease Control and Prevention, the U.S. Department of Agriculture, and the academic community. Not surprisingly, similar interests exist among these agencies for rapid, accurate, and complete identification of the etiologic agent as well as the desire to identify the source of a bioweapon with as much uniqueness and statistical confidence as possible. This naturally leads to thinking about ways to maximally achieve the phenotypic and genotypic discrimination of microorganisms well beyond strain or isolate. Earlier discussion in this volume on the genetic variability of HIV is instructive. Jenks's 1998 commentary on microbial heterogeneity (including the presence of "orphans," reminiscent of noncoding regions used for forensics purposes with human samples) also is useful. A focus on core diagnostic technology alone is insufficient. To achieve success, this must be part of an end-to-end system with several components fitted together and a range of issues and equities considered.

A national end-to-end architecture should be established that enables the coherent, efficient, fully exploitative, and legally and politically defensible capability to collect, preserve, analyze, and interpret physical and biological evidence associated with an illicit biological event. A consortium consisting of law enforcement, defense, intelligence, public health, and agriculture agency laboratories could be formed to design, operate, and manage this system. Through this consortium, relevant capabilities could be identified, coordinated, integrated, and optimized in a systems approach. This proposed system, or portions thereof, would address not only evidence collected after the fact but also "evidence" collected before the fact. (Reference samples as isolates of infectious disease microorganisms would fall into the latter category.)

All biological and other physical evidence would be properly collected and preserved and its "chain-of-custody" documented in accordance with accepted forensics and legal requirements. Rapid preliminary diagnostics would be accommodated in a network of specially certified public health or medical (Centers for Disease Control and Prevention, 1999; McDade, 1999) and veterinary or plant pathology diagnostic laboratories. Confirmatory identifications and in-depth characterizations would be conducted in designated bioforensics laboratories.

In addition, a limited number of secure government biocontainment facilities would be modified to allow for safe and effective forensics analysis of contaminated physical evidence by qualified, certified, and specially trained forensics specialists. Evidence handling and examination practices by and between all laboratories in this system would be in accordance with accepted forensics practices and legal requirements to incorporate chain-of-custody practices as well as prevent the alteration, cross-contamination, or deleterious change of any evidence. Provisions for subsequent examination by the defense may be required. Secure facilities would have to be constructed for the safe storage of contaminated probative evidence. This could conceivably mean that hundreds or thousands of exhibits would have to be maintained, possibly for years, until the adjudication occurs. Quality assurance standards, which would ensure accuracy, consistency, validity, and confidence in the practitioners, technologies, methods, and results should be instituted and audited across this enhanced bioforensics laboratory system. Research funding should be appropriated to develop, validate, and advance forensics capabilities with biological as well as associated forms of WMD evidence.

CRITICAL COMPONENTS OF A BIOFORENSICS CAPABILITY

The ability to maximally attribute an organism of interest to a source with a very high degree of scientific certainty is a keystone of this proposed architecture. Bioforensics reaches its full potential and impact when we design, build, and integrate several critical components. We must begin with a set of validated, highly informative, and discriminating approaches to characterizing microbial genotypes and perhaps phenotypes. Further development of the model occurs by drawing from what already exists in the United States in forensics (human) DNA analysis, including through the national forensics DNA informatics system known as CODIS (Combined DNA Indexing System) and the well-described high-throughput or batch science laboratories proposed by Layne and Beugelsdijk (1998, 1999).

To meet anticipated demand, the subsystem for bioforensics would require automated microbial analysis systems that are fully validated for biomedical, forensics, and agricultural samples and purposes. Large computerized reference libraries would need to be created or interfaced that contain all of the pertinent data on all of the microbes of interest, including all isolates known and newly collected. This implies the need to accommodate a constant flow of samples into the system and a surge capacity in response to a bioevent or when an investigation is undertaken. Interconnectivity and interoperability between databases is a requirement. At the same time, full data security and integrity must be maintained. Needed would be informatics that permit full and rapid description and discrimination between microorganisms from known and questioned sources, using perhaps both phenotypic and genotypic data. Finally, statistical approaches would be required that quantify and communicate the power and confidence of the identification (or exclusion).

As noted, there is a corollary for this in forensics science (Kirby, 1993). Beginning in 1989 the FBI envisioned a DNA informatics system that embodies several of these features. CODIS was first deployed in 1992. Today, it is used in more than 110 forensics laboratories in the United States and increasingly in international forensics laboratories. CODIS permits law enforcement agencies to share key investigative information, such as DNA profiles, that can link and focus investigations and facilitate the rapid identification of perpetrators. Since its inception, CODIS has facilitated the solution of violent crimes, whether committed recently or years ago and whether they involved neighboring or widely separated jurisdictions.

One poignant example came to light in 1999. Approximately 9 years earlier a number of women were murdered in the Oklahoma City, Oklahoma, area. Bloodstain evidence in those cases was submitted to the Oklahoma Bureau of Investigation (OBI) Laboratory. The cases remained unsolved until, while reviewing this case, the OBI provided DNA profiles to be uploaded into various state CODIS databases. A "hit" (match) with a known offender profile was obtained from the California CODIS database almost immediately. Through this a suspect was identified, whose specific whereabouts were at that time unknown. The FBI was contacted and very quickly located the suspect and arrested him, all in a matter of days. Attribution, via biological evidence and forensics DNA technology, was key to providing direction to the investigators in resolving this matter.

Since its conception, CODIS has evolved significantly. Today, U.S. CODIS laboratories are connected through a high-speed telecommunications network, known as the CJISWAN, which is owned and operated by the FBI. By convention, CODIS accommodates standard sets of DNA loci (RFLP, STR, PCR, and mtDNA) and fundamentally contains two types of DNA information. All 50 states permit the collection, analysis, and submission to CODIS of DNA profiles from known samples from individuals convicted of violent felonies. (The FBI is awaiting passage of federal legislation to permit the collection, analysis, and uploading of federal convicted-offender samples.) Known convicted-offender samples are entered into CODIS as analyzed by state, local, or contract laboratories. These data, in fact, form a system of interconnected "libraries" against which samples of unknown origin are compared. Samples from unidentified sources, such as human remains or John Does can also be submitted in an attempt to identify the donors. CODIS is arranged in a hierarchy of local, state, and national layers and uses a protocol for access and use of the information. Legal requirements, such as privacy protections, have been incorporated and are meticulously enforced. CODIS laboratories are regularly audited from several perspectives. New search engines are being developed at present to accommodate faster response for rapidly expanding databases. The impact of bioforensics and beyond to other domains of WMD forensics can easily be envisioned.

A national forensics capability can be established in the near term, which would provide a powerful tool for the investigation and prosecution of the possession or use of biological weapons, whether through acts of terrorism, proliferation, or even war. Several key components and characteristics have been identified in this paper, with high-throughput laboratories prominent among them. The approach suggested here encourages convergence and integration of a number of different fields, expertise, technologies, and interests toward effectively dealing with a national prioritized threat. The relationship to infectious disease surveillance is also evident. A relatively small investment, perhaps in the range of a few tens of millions of dollars, could provide a capability with far-reaching and even global implications.

SUMMARY

This paper has concentrated on the role of forensics in attributing acts of bioterrorism or bioproliferation and has suggested that national capabilities could be enhanced considerably by establishing a laboratory network that includes advanced diagnostics, high-throughput robotics, and powerful informatics interconnected by a high-speed telecommunications backbone. This system would be supported by robust evidence management processes and complemented by advanced traditional forensics that would permit full exploitation of any physical evidence associated with these acts. Finally, I present this question for your consideration: Would a powerful bioforensics capability also contribute to deterring the use of biological weapons?

REFERENCES

Alibek, K., and S. Handelman. 1999. Biohazard. New York: Random House.

Anderson, J. H. 1998. Microbes and Mass Casualties: Defending America Against Bioterrorism. Washington, D.C.: The Heritage Foundation.

Bardi, J. 1999. Aftermath of a hypothetical smallpox disaster. Emerging Infectious Diseases (www.cdc.gov/NCIDOD/EID/vol5no4/bardi.htm).

Bartlett, J. G. 1999. Applying lessons learned from anthrax case history to other scenarios. Emerging Infectious Diseases (www.cdc.gov/NCIDOD/EID/vol5no4/bartlett.htm).

Centers for Disease Control and Prevention. 1999. Preventing Emerging Infectious Diseases: A Strategy for the 21st Century (www.cdc.gov/NCIDOD/emergplan/index.htm).

Clarke, R. A. 1999. Finding the right balance against bioterrorism. Emerging Infectious Diseases (*www.cdc.gov/NCIDOD/EID/vol5no4/clarke.htm*).

Commission to Assess the Organization of the Federal Government to Combat the Proliferation of Weapons of Mass Destruction. 1999. Combating Proliferation of Weapons of Mass Destruction. Washington, D.C.: U.S. Government Printing Office.

Ember, L. 1996. Science center to handle terrorism at Olympics. Chemical and Engineering News, 74(29):11-12.

Falkenrath, R. A., R. D. Newman, and B. A. Thayer. 1998. America's Achilles' Heel: Nuclear, Biological and Chemical Terrorism and Covert Attack. Belfer Center for Science and International Affairs, Harvard University, Cambridge, Mass.

Fox, J. 1998. Bioterrorism: Microbiology key to dealing with threats. American Society for Microbiology News, 64(5):255-257.

Franz, D. R. 1999. Defense against toxin weapons. The DTIC Review, 4(3):1-51.

Hamburg, M. A. 1999. Addressing bioterrorist threats: Where do we go from here? Emerging Infectious Diseases (www.cdc.gov/NCIDOD/EID/vol5no4/hamburg.htm).

Henderson, D. A. 1999. The looming threat of bioterrorism. Science, 283(5406):1279-1282.

Hughes, J. M. 1999. The emerging threat of bioterrorism. Emerging Infectious Diseases (www.cdc.gov/NCIDOD/EID/vol5no4/hughes.htm).

Inglesby, T. V. 1999. Anthrax: A possible case history. Emerging Infectious Diseases (www.cdc.gov/NCIDOD/EID/vol5no4/inglesby.htm).

Jenks, P. J. 1998. Sequencing microbial genomes: What will it do for microbiology? Journal of Medical Microbiology, 47:375-382.

Kaplan, D. 1998. Cult at the End of the World. New York: Random House.

Kirby, L. T. 1993. The Combined DNA Indexing System (CODIS): A theoretical model. In DNA Fingerprinting, A Practical Introduction. New York: W. H. Freeman.

Kortepeter, M. G., and G. W. Parker. 1999. Potential biological weapons threats. Emerging Infectious Diseases (www.cdc.gov/NCIDOD/EID/kortepeter.htm).

Layne, S. P., and T. J. Beugelsdijk. 1998. Mass customized testing and manufacturing via the Internet. Robotics and Computer-Integrated Manufacturing, 14:377-387.

Layne, S. P., and T. J. Beugelsdijk. 1999. Laboratory firepower for infectious disease research. Nature Biotechnology, 16:825-829.

MacKenzie, D. 1998. Bioarmegeddon. New Scientist, 42-46.

McDade, J. E. 1999. Addressing the potential threat of bioterrorism: Value added to an improved public health infrastructure. Emerging Infectious Diseasès (www.cdc.gov/NCIDOD/EID/vol5no4/mcdade.htm).

O'Toole, T. 1999. Smallpox: An attack scenario. Emerging Infectious Diseases (www.cdc.gov/NCIDOD/EID/vol5no4/otoole.htm).

Pavlin, J. A. 1999. Epidemiology of bioterrorism. Emerging Infectious Diseases (www.cdc.gov/NCIDOD/EID/vol5no4/pavlin.htm).

Rimmington, A. 1999. Anti-Livestock and Anti-Crop Offensive Biological Warfare Programmes in Russia and the Newly Independent Republics. Report from the Centre for Russian and East European Studies, University of Birmingham, England.

Siegrist, D. W. 1999. The threat of biological attack: Why concern now? Emerging Infectious Diseases (www.cdc.gov/NCIDOD/EID/vol5no4/siegrist.htm).

Snyder, J. W. 1999. Responding to bioterrorism: The role of the microbiology laboratory. American Society of Microbiology News, 65(8):524-525.

Stern, J. 1999. The prospect of domestic bioterrorism. Emerging Infectious Diseases (www.cdc.gov/NCIDOD/EID/vol5no4/stern.htm).

Tucker, J. B. 1999. Bioterrorism is the least of our worries. New York Times, October 16.

19

Biological Warfare Scenarios

William Patrick III

INTRODUCTION

This paper will discuss two vulnerability tests, neither of which could be talked about in an open forum until 1999, when the information became public. One of these was a large-scale aerosol test that demonstrated the vulnerability of a seaport, San Francisco. This test took place in 1950. The second test, conducted in 1965, was a simulated attack on an enclosed environment, the subway system of New York City.

LINE SOURCE DISSEMINATION

Today, we tend to talk about biological warfare agents in great detail and ignore the impact of the munitions system, the delivery system, and the meteorological conditions at the target. In addition, a number of important parameters exist that a would-be terrorist must address in order to be successful. Although a discussion of "weaponization 101," as illustrated in Figure 19.1, is beyond the scope of this paper, it is important to discuss the line source dissemination (see Figure 19.2). Line source could be accomplished by a high-performance aircraft, or it could be an individual walking along a line with a 2-gallon spray tank disseminating a liquid perpendicular to the wind, with the energy of the wind taking the aerosol downwind. This is by far the most effective way to deliver a biological warfare agent. Fortunately, line source is very susceptible to meteorological conditions, such as changing winds.

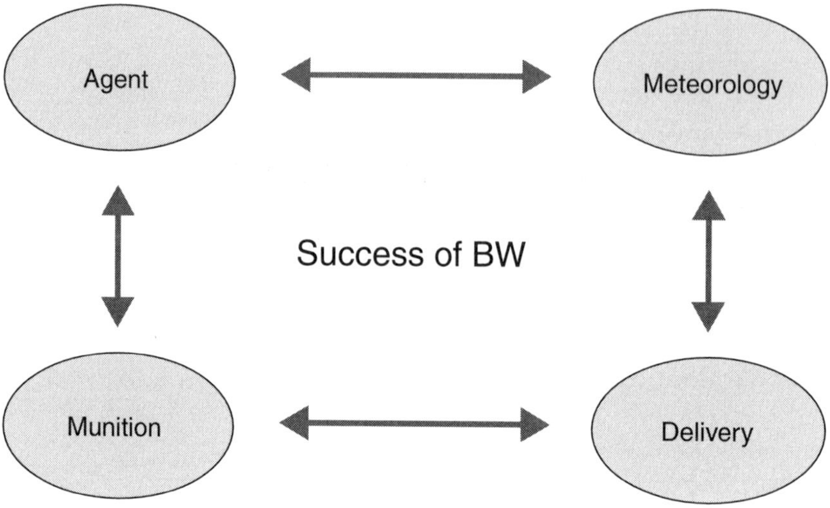

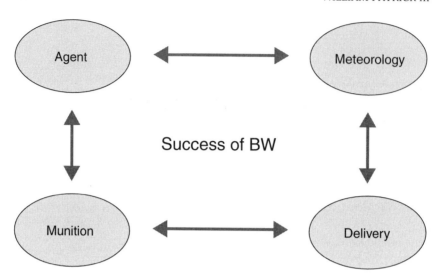

FIGURE 19.1 Four components which must be addressed in an offensive biological warfare program.

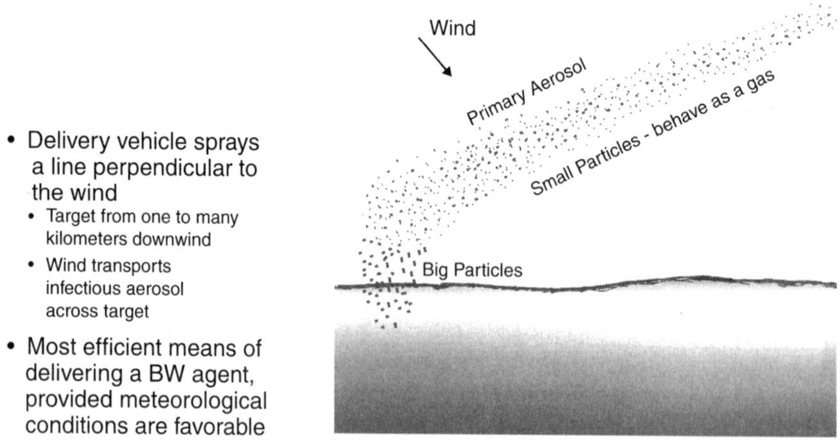

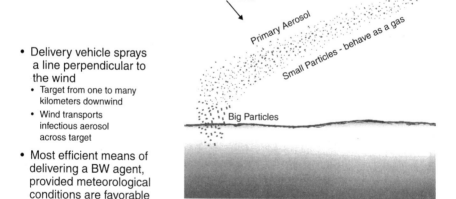

FIGURE 19.2 Line source dissemination.

Regarding the nature of the aerosol, it is important to note that there is a period of time immediately following dissemination (whether from an aircraft or any other disseminating device) when the aerosol comes into equilibrium with atmospheric conditions. During this period, referred to as the time of equilibration, the big particles fall out of the aerosol, land on the terrain, and form strong adhesive bonds with the surface (see Figure 19.3).

It is extremely difficult to get these particles to reaerosolize, which is called a secondary aerosol. However, it is the primary aerosol, which is composed of particles within the magic size range of 1 to 5 microns and which behaves as a gas, that remains airborne and causes infections. Once the primary aerosol is formed, these small particles remain airborne. This is one of the major differences between a chemical attack and a biological warfare attack; large quantities of decon are not needed to treat the area over which aerosol passes. Small particles do not fall out. Infections occur because we act as vacuum pumps, pulling in the small particles.

When a helicopter lands in an area that has been potentially contaminated by the fallout of organisms from a primary aerosol (which is very, very low), there will be little or no contamination on the helicopter or the personnel in that area, reflecting the fact that primary aerosols are composed of small particles that remain airborne. Problems arise when the ground is deliberately sprayed directly with either powder or liquid containing microorganisms. As a general rule, the concentration of organisms on the ground must exceed 1×10^7 cells per meter square. This was

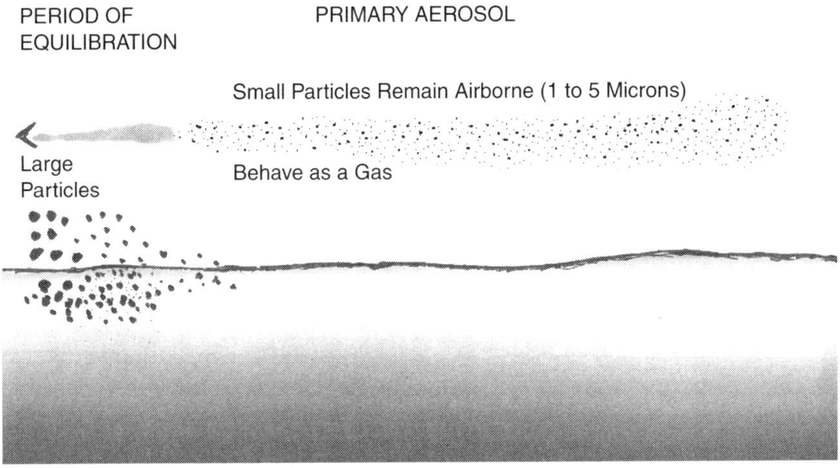

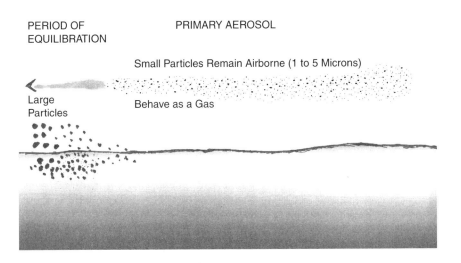

FIGURE 19.3 Physics of primary aerosol.

established in early Detrick studies (in the 1950s) and later by a biophysicist at Dugway Proving Ground (35 to 40 years later). Concentrations of 1×10^3, 10^4, and 10^5 organisms per square meter do not easily produce secondary aerosols unless high levels of energy are applied. These adhesive bonds are so strong that a high concentration of organism must be reached in order to overcome the bonds between organism and terrain.

The nature of the primary aerosol can be demonstrated effectively by tests conducted in the Pacific and the Arctic. The harmless simulant *Bacillus globigii* (BG) was disseminated in a series of tests demonstrating the vulnerability of naval vessels to small-particle primary aerosols. The aerosol is pulled into the ship by the air system and remains in a high concentration for about 1.5 to 2 hours. Then it departs the ship, leaving very little residue. The primary aerosol, which behaves as a gas, deposits little or no BG spore on the floors and walls of the ship, although the concentration of spores is high in the air. After the aerosol passed through the ship, the floors and doors were swabbed and cultures prepared; however, little contamination was found. In fact, the level of contamination was so low that a small quantity of seawater effectively removed it.

In conducting these simulant tests, the aerosols were sampled with the all-glass impinger (see Figure 19.4). Hundreds of these samplers were used during an open-air test. The impinger contains a fluid—the impinger fluid. Air is pulled in at a specified rate of 10 liters per minute (roughly the breathing rate of a man at rest), due to a calibrated orifice on the outtake. If the material collected in the sampler is determined and the length of time the impinger has operated is known, a relatively accurate fix can be made on the number of spores that would be inhaled by an individual.

SAN FRANCISCO

Using hundreds of these impinger samplers, a vulnerability test was conducted in 1950 on the city of San Francisco using the simulant BG. The purpose of the test was simply to determine if a seaport is vulnerable to biological warfare attack by line source dissemination.

A small naval vessel sprayed a line 2 miles long 2 miles off shore just at sundown using liquid BG (see Figure 19.5). In the first test a strong inversion was present with a gentle wind of about 10 mph. The downtown area of San Francisco was heavily contaminated with samplers, indicating more than 10,000 spores per liter. This concentration would have caused more than 60 percent infections among the population. When an aerosol is released, even under the best of conditions it is difficult to predict where it will go. Thus, the Berkeley area was also contaminated, although at a much lower level. We concluded that this test was highly successful.

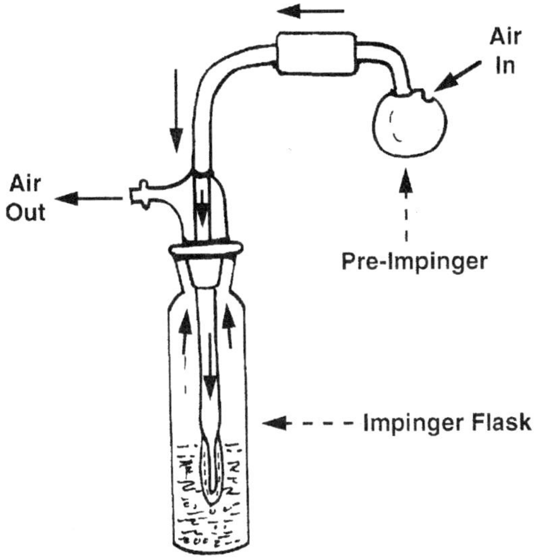
Air
In
Air
Out
Pre-Impinger
Impinger Flask

FIGURE 19.4 All glass impinger with pre-impinger.

The next test was conducted in an unstable air mass. Again, BG slurry was disseminated as a line 2 miles long. The same amount of BG was used. A high concentration of spores was found only about two blocks into the city. An unstable air mass failed to achieve the projected casualties, demonstrating that meteorological conditions on an open target are

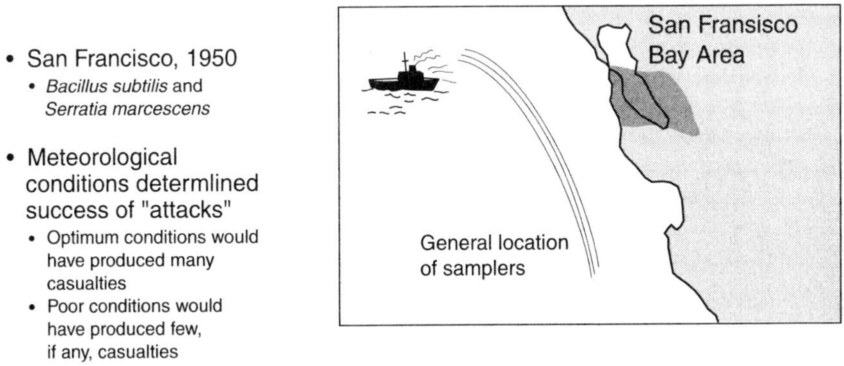
• San Francisco, 1950
• Bacillus subtilis and Serratia marcescens
• Meteorological conditions determlined success of "attacks"
• Optimum conditions would have produced many casualties
• Poor conditions would have produced few, if any, casualties
San Fransisco
Bay Area
General location
of samplers

FIGURE 19.5 Simulated attacks.

as important to the success of an attack as the agent, the munition, and the disseminating system.

A vegetative organism, *Serratia marcescens*, was also tested for its ability to contaminate San Francisco. The line of dissemination and general test conditions were similar to those for BG. The dissemination was made under good meteorological conditions, including moderate inversion and a 12 mph wind. The impingers indicated that only about 25 cells were recovered per liter in the first blocks of the city, suggesting that the attack was a failure. Upon subsequent testing we learned that, even though this test had been conducted at sundown, sufficient ultraviolet light was present to kill a vegetative cell. These tests were conducted by a well-supported program that included microbiologists, aerobiologists, meteorologists, and munitions development engineers. The results should be expected to be good if this type of support is available and meteorological conditions are favorable.

DISSEMINATION ISSUES

One of the principles of biological warfare that we learned from our former offensive program is that, although liquid agents are relatively easy to make, they are very difficult to disseminate into a small-particle aerosol. A single-fluid nozzle with gaseous energy is one of the simplest ways to disseminate a biological warfare agent; however, it is not very efficient. Most of the particles are large and fall out of the aerosol quickly. For a single-fluid nozzle to achieve a 5 percent level of efficiency, it would require a minimum of 300 psi. At this level of pressure the container would have to be made of metal, not glass or plastic. The would-be terrorist must not only produce the agent but also requires a model shop in order to construct the agent container and combine it with an appropriate nozzle.

We have discussed big particles that fall out of the aerosol quite quickly. For biological warfare to be successful, a primary aerosol that is composed of 1- to 5-micron particles must be generated. The classic Ft. Detrick experiment compares man to the monkey and guinea pig. The volunteers for this study came from the Seventh Day Adventist Church. It must be emphasized that, at the time this study was conducted, the Soviet Union and Red China were our enemies, and although young people from the Seventh Day Adventist Church wanted to serve their country, they did not want to carry rifles. Therefore, they instead volunteered to be exposed to a series of organisms. The first aerosol tests were conducted with *Coxiella burnetti*, the causative agent of Q fever. Two years later *Franciscella tularensis* was used, and still later Staphylococcal enterotoxin B was tested. These diseases are self-limiting and can be effectively treated with antibiotics. The volunteers recovered and have been followed medically over the years, with no adverse responses noted.

These were very important studies because when the volunteers were exposed, rhesus monkeys and guinea pigs were also exposed. Thus, a relationship was developed between the human and animal models that could then be applied to other diseases for which people could not have been exposed to testing for ethical reasons.

The first column of Table 19.1 shows aerosol particle size. The second column demonstrates the number of cells of tularemia required to kill the guinea pig at the 50 percent level, a respiratory LD_{50}. The third column illustrates the number of tularemia cells needed to kill the monkey. The last column shows the number of cells of tularemia required to infect but not kill man (an ID_{50}). An aerosol composed of 1-micron particles of tularemia requires only 2.5 cells to kill the guinea pig, 14 for the monkey, and between 10 and 52 cells to infect man. If the tularemia culture is less than 48 hours, the infecting dose for man is between one and 10 cells. This is the limit of assay precision. However, delivering a culture within 48 hours of its production is not operationally feasible.

When the aerosol is composed of 6.5-micron particles, a larger number of cells are now required to infect by the respiratory route. When the aerosol is composed of 18- to 22-micron particles, the number of tularemia cells becomes extremely large. Man was not exposed to these large particles because of other more important studies such as vaccine efficiency. This experiment clearly demonstrates that the biological warfare agent must be disseminated into a small-particle aerosol.

RELIGIOUS CULTS AND BIOTERRORISM

I would like to address the problem of religious cults and bioterrorism, specifically, the Aum Shrinrikyo cult in Japan. Two investigative

TABLE 19.1 Classic Experiment: Man-Monkey-Guinea Pig: Influences of Particle Size on Tularemia Infectivity

Aerosol Particle Diameter (Microns)	Guinea Pig RLD_{50}	Monkey RLD_{50}	Man RID_{50}
1	2.5	14	10 – 52
6.5	4,700	178	14 – 162
11.5	23,000	672	No Data
18	125,000	3,447	No Data
22	230,000	> 8,500	No Data

reporters for the *New York Times*, Miller and Broad, learned from various sources that this well-funded cult had allegedly disseminated liquid anthrax cultures on perhaps as many as nine occasions. All of these attacks failed to produce a single infection, and at the time the attacks were not even detected. Why did the Aum Shrinrikyo fail when the organization had modern laboratories, trained personnel, and sufficient funds? I believe the Aum failed because it did not meet the four essential components presented in Figure 19.1 for a successful biological warfare attack. First, it may not have selected a virulent strain of anthrax. Selection of the virulent strain is the most important in agent weaponization. For example, during the U.S. offensive program, many strains of anthrax were studied before selecting the most appropriate one for weaponization. Next, the munitions and delivery systems may not have been appropriate. Finally, it ignored meteorological conditions. One attack supposedly occurred from an eight-story building, at midday, in downtown Tokyo.

It is important to remember that liquid cultures are difficult to disseminate into small particles, and the disseminating device or munitions requires high levels of energy for success. Table 19.2 illustrates how agent viability interacts with particle size. Dry *Serratia marcescens* or SM is a very small vegetative cell. If the aerosol contains particulates in the 0.8-micron range, there are only 1.8 cells on average in these particles and the viability is 0.001 percent. As particle size increases, the viability of cells in the aerosol particle increases. Thus, big aerosol particulates contain viable cells, yet small particles are most effective in causing a respiratory infection. Therefore, our would-be terrorist has some basic problems that require solutions. (We cannot discuss in an open forum how this and similar problems are solved.) I suspect that biological warfare may have been attempted in this country and failed, and failure is not usually advertised.

TABLE 19-2 Dry Serratia Marcescans (SM): Relationship of Particle Size, Viable Cells per Particle, and Viable Cells per 1,000 Particles

Aerosol Particle Size	SM per Aerosol Particulate	Viable SM Cells per Aerosol Particulate	Viable SM Cells Frequency per 1,000 Aerosol Particulate
0.8	1.8	0.001	0.5
1.3	4.2	0.01	2.6
3.0	18.0	0.2	15.6
6.5	73.0	2.5	38.0
11.5	195.0	7.7	14.0
16.0	350.0	11.0	60.0

NEW YORK CITY SUBWAY SYSTEM

The next scenario involves an enclosed environment where meteorological conditions are no longer a factor. One of the most important vulnerability studies conducted during our offensive program involved the New York City subway system. A simulant powder containing BG was prepared that possessed very good secondary aerosol properties. Lightbulbs were filled with BG and were dropped from the back of trains onto the subway tracks. Impinger samplers had been distributed throughout the subway system to include both trains and stations.

The passage of the trains over the powder created secondary aerosols that were carried throughout the entire subway system. The BG penetrated all test trains and remained in high concentration for 1 to 1.5 hours. Thereafter, with the dilution factor at work, the concentration dropped markedly, and after 2 hours the impinger samplers were not yielding spores. The risk of infection if a biological warfare agent had been used would have been highest for personnel using the subway near the site of powder drop and within the first hour following dissemination.

Studies have shown that in 1965 the average time people spent on the trains during rush hour (morning and afternoon) was about 8 minutes. Thus, impinger data that determined the number of organisms per liter of air and the number of minutes that people were on a train indicated that about 80 to 90 percent of the train population would have become infected. (If treatment is not started early in the disease process when the first subtle symptoms appear, anthrax is fatal.)

CONCLUSION

At this time, domestic terrorists do not have the capability to develop a biological warfare weapon that would result in serious casualties. However, there is concern that a state-supported group with trained personnel and adequate laboratories and funds could develop an agent powder with the appropriate biological and physical properties and that a few hundred grams of this powder, which could enter the United States via those with diplomatic immunity, if used in an enclosed environment, would produce thousands of casualties. The questions then become: How does the United States determine the perpetrator? and What is the response?

PART IV

FURTHER APPLICATIONS AND TECHNOLOGIES

20

Integration of New Technologies in the Future of the Biological Sciences

David J. Galas and T. Gregory Dewey

INTRODUCTION

Integration of technologies into the life sciences, which is critical to their advance, will take two forms. The first is the integration of existing technologies to produce more powerful tools for discovery. The second is the integration of technological change with discovery as an integral part of the process of scientific advancement. This paper discusses and illustrates both of these meanings, taking our lead from the biology of the past few decades, which have seen science progress from the foundations of molecular biology to the high-throughput parallel data acquisition of genomic and other data that characterize the biology of the past few years. The applications that are focused on the understanding and control of infectious diseases and biological aggression are specific instances of a much broader area of application of these ideas. The principal conclusions of this discussion are that advances in physical technologies, in particular automation and miniaturization, and the increasingly interdisciplinary nature of technological development are vital to the future of the life sciences. This is largely because the inherent complexity of biological systems requires a massive increase in an already high rate of data acquisition. Analysis of these data and their synthesis into new scientific knowledge will dominate the foreseeable future of the biological sciences. The essential scientific and technological activity required to meet this goal of data assimilation is the development of a modeling capability for complex biological systems that greatly exceeds our present capabilities. We discuss prospects and constraints on the development of this interface between theory and experiment in the biological sciences. Finally, since the interfaces between disciplines and the combinations of scientific and engineering points of view in the biological sciences are new, the cross-disciplinary education and training of a new breed of professionals in the applied life sciences are crucial to achieving this future.

DISCOVERY AND TECHNOLOGICAL CHANGE

The life sciences are unique among the sciences in several ways. One of these is that the history of the biological sciences has been almost entirely driven by the invention of new technologies, from van Leeuwenhoek's first microscope, which opened the world of the very small to human observation, to the invention of DNA cloning technology in the 1970s, which opened the world of the biomacromolecule to human experimentation. Access to realms of the hitherto unknown provided by each successive invention has marked the advance of biological discovery. Discovery has marched relentlessly in the footsteps of technological innovation. The pattern has been a repeated cycle of new technologies opening the door to discovery and then new discovery in its turn enabling or motivating the development of newer technologies. The history of this inexorable technology-discovery cycle in biology is a long one that includes a wide range of advances. Perhaps most relevant to our subject are the technological inventions that can be credited with creating modern biology in the last half of this century. Our list of these is an arguable one by all accounts, but many of our entries would likely be included by most.

First, most would agree that modern biology has been ushered in by that bastard science born of chemistry, physics, and biology called molecular biology—a classically feeble term that attempts to describe the field. It was roundly criticized at its inception by eminent scientists such as biochemist Erwin Chargaff as a superficial integration of sciences and a shallow discipline, one not sufficiently biochemical or biological to be taken seriously (Cairns et al., 1966; Watson, 1968). Yet it proved to be the key to unlocking the advances in biological sciences that marked the last half of the twentieth century. The list must include Cohen's and Boyer's methods for cloning DNA molecules (Cohen et al., 1973; Boyer, 1971) and discovery of the tools that make such cloning possible, restriction enzymes (Nathans and Smith, 1975) and DNA ligase; the sequence-specific hybridization of DNA sequences as a way of identifying specific sequences (Southern, 1974); monoclonal antibody methods (Köhler and Milstein, 1975); methods for deciphering the three-dimensional structure of macromolecules; invention of DNA sequencing methods and the subsequent technical elevation of one of these methods to the automated DNA sequencing machine (Sanger and Coulson, 1975; Maxam and Gilbert, 1977);

methods for direct chemical synthesis of DNA (Itakura et al., 1984); and the method for the amplification of nucleic acid sequences by more than a millionfold—the polymerase chain reaction (Arnheim and Erlich, 1992).

These technologies and their descendants have enabled the partially completed dissection of the molecular components and interaction of the living cell and continue to transform modern biology. It is becoming clear, however, that we are on the brink of another revolution that is characterized by the convergence of several technical and scientific developments. We argue here that the present change will have as great an impact as the change that brought forth molecular biology almost 50 years ago.

FUTURE REALM OF TECHNOLOGY AND DISCOVERY

At the threshold of the twenty-first century, the pace of technological innovation in the life sciences seems to be increasing and promises to carry the science of biology into a new realm in a very few years. In the past few years a technological phenomenon has been at work in another realm: computer and communications technologies have transformed many areas of everyday life and business. Inevitably, the life sciences have also been affected. The combination of high-throughput data acquisition, epitomized perhaps by the current final push to sequence the human genome, and the relentless surge of computing and data storage power is about to bring forth a new realm in the science of biology. To underestimate the effect of what some would consider a protean sector on the fundamental understanding of biology would be a grave error.

The realm we are thinking of is one that can only be projected from the promise of developments in the present state of the science and the technology. This future is very different from even the past decade in biology and will be bristling with new technologies for measurement and manipulation, flooded with data and information about complex systems operating in biological cells and organisms, and empowered with methods for directly modifying cells and organisms. But most of all this data-rich realm of molecular detail has become completely quantitative and is now described in terms of mathematical equations and complex computational models. The genome sequencing projects of the past have provided the essential raw data that have finally enabled the compiling of accurate molecular "parts lists" for living cells that are the central information resources for biological research.

The methods of acquiring information and data are now very advanced and continue to develop rapidly. They have become fully automated and high throughput in character, with many tasks being carried out simultaneously. Much of the acceleration and enhancement of data acquisition rates in these new technologies have been achieved through miniaturization to micron scales and a high degree of intelligent automation—experimental protocols in which many of the measurement decisions, sample tracking, and data storage and analysis are now entirely computer directed.

Needless to say, laboratory instrumentation has finally been organized into local networks of flexibly controlled tools that can be guided by a central interface that aids researchers with intelligent assessments of the tasks and commands for a project. Laboratory automation is now very sophisticated and well advanced. Data flow directly to the software agents for analysis and perusal and final presentation to the experimenter. The deciphering of complex biological systems—understanding how they work—has now taken on an entirely new character, unexpected to some. Hypotheses about how a system works are no longer described by biologists in words (e.g., protein A represses the function and expression of the gene for protein B, or protein A interacts with receptor B and stimulates the cell to divide), but rather they must be embodied in mathematical models defining and integrating the interactions and dynamics of the whole system in a quantitative and prediction-generating fashion.

This transition to mathematical models has occurred largely because of the complete inadequacy of the descriptive mode in handling complex systems that have multiple nonlinear interactions and responses. The accumulation of overwhelming evidence of the inherently complex nature of the systems has forced entirely new methods of analysis. In the year 2000 we are just beginning to face the fact that the process of constructing large quantitative models from large complex datasets was something we did not know how to do. Conversely, we did not know how to critically compare large datasets with existing models. We had no idea how to obtain values for all the molecular parameters or how to measure the rate constants and interaction strengths that went into the models. We also did not know how accurate they had to be to produce a useful model. Somehow these problems were solved with the usual combination of surprises, disappointments, and triumphs.

The transition of molecular biology from molecular natural history—descriptive lists of components and classifications—to real biological science with predictive models of behavior has been largely completed. Researchers have now come to grips with the hard lessons learned from the 1970s through the 1990s, a period during which most experimental studies were interpreted in a strictly descriptive mode. This necessary phase of scientific evolution has passed. Predictions were then made on the basis of these qualitative, poorly formed models that often resembled a conceptual Rube Goldberg machine. These linear chains of causation were weakened by many assumptions and often had poor or limited predictive value, giving a prediction that in some cases was found to be different from expectations—even the opposite of the experimental findings. Since the molecular systems in question likely have significant nonlinear characteristics, this was not a surprising outcome.

There is now an enormous spectrum of models and classes of models that are based on cooperative collective phenomena. Although these models treat the system as a whole, with the new computational power they can be relatively easily manipulated, used on past data, and tested and validated using the software available. In short, the transition of biology into a mature science, with a theoretical component to guide its development and a robust experimental infrastructure to provide data, has finally occurred, and the science and its applications are thriving.

The lessons of the past decades in constructing and maintaining massive and complex databases have also been incorporated into the present network of interconnected and relatively transparent sets of databases distributed throughout the reach of computer networks. Publicly and privately funded databases that are centrally assembled and maintained, dispersed private researcher-assembled special databases, and company databases, open and closed, are all now connected and accessible. These databases can be seamlessly connected in complex queries and transposed from one operating system to another (Liu et al., 2000).

Principal among these lessons is that the way data or information is displayed and the way the data structures within the databases are defined have a strong influence on how the data can and will be used by scientists. This, in turn, influences the inferences that can be made from the data. To perceive all the regularities and see the patterns and trends, however subtle, the data must be appropriately displayed so that the right questions are asked. Thus, there is a need for flexible and easily reconfigured data structures that can be redesigned on the fly by users to fit their specific viewpoints and new evolving models. Such data structures are in the ascendant.

We have gone well beyond relational databases and are now in an era of higher-order, object-oriented databases. These databases store information in the form of "objects" as well as functions specific to the databases. The functions are tailored to the potential applications of the objects, and the entire database has qualities resembling higher-level simulations of the previous era. The databases, in fact, mimic both the components and the functionality of the biological subsystems they represent. With the new functionality of these databases, a variety of problems are easily handled; even the aggravating and persistent problem of inconsistent data is mitigated by "self-proofing" software that seeks out ever more subtle inconsistencies between datasets and calls them to the attention of the researcher. Database languages and the logic of simulations have converged to a large extent.

It is now only very rarely that one can capture a precise and useful hypothesis sufficiently in a brief description. Most working models of the cell and organism have become complex computer codes whose simulations are compared with experiments to refine parameters such as binding constants and reaction rates, to define new interactions or new components, or to test the response of the system to a new set of stimuli. These codes are further tested by experimental directed modification of the cellular components, through genomic modifications, or by highly specific chemical modifications.

These biocomputing codes are similar to the large physical codes that model the behavior of the atmosphere and oceans of our planet and provide a strong, if somewhat short-range, predictive capability. While it took more than 30 years to move from the first crude climate modeling codes to the first realistic and predictive codes, development of the biological equivalents has gone much more quickly, partly as a result of this experience and partly from the much more powerful computer technology. The complexity of these codes is now built up using recursive object-oriented designs. We have learned which subsystems in an organism can be modularized in this manner. Higher-order organizations are then constructed by assembly of computational modules. This allows a stepwise self-consistent increase in complexity of the collective model. This fits well with real biological systems, metaphorically recapitulating the possible evolutionary pathways of the biological systems.

It is now possible to assemble sets of genes from a variety of organisms into actual cellular enclosures (stocked with the essential subcellular components), to observe the operation of these cellular (and multicellular) systems, and to compare their functioning with that predicted by the appropriate computer code. These "cell-sims" are teaching us new things every day about the emergent behavior inherent in a sufficiently complex set of genes and about the complex computer codes we now use to simulate them.

Real engineering in biology has now emerged as a powerful computationally driven technology, and many of the lessons of the past transitions of disciplines into the "engineering phase" are being applied to biology. In this realm of the future we have been able to devise completely unexpected mechanisms of cellular response by observing this kind of collective behavior and modifying it according to the predictions of an iteratively modified computer code. We are now beginning to devise methods for efficiently modifying these modeling codes so that precisely defined cellular behavior can be predicted and built into the cells. Engineering at this level has begun to resemble in some ways the computational design of integrated circuits at the end of the twentieth century. It was possible in the latter part of the century to redirect the designer's attention from the individual transistor, diode, or gate to design on a larger scale of thousands of these circuit elements at once using computer tools. The development of "cell-sims" has a great advantage over integrated circuits. They have the potential to evolve. High-throughput screening coupled with directed evolution allows a massive exploration of functional motifs on a molecular and systems level. Molecular evolution in the test tube, initially explored in the 1980s (Beaudry and Joyce, 1992), has become a powerful molecular technology at the boundary of synthetic chemistry and computing. Molecular structures can be evolved in vitro in a few hours to meet the requirements of very complex criteria of properties and behavior.

New professions have arisen in the area of design, construction, use, and maintenance of these computer code sims, analogous to the chip designers and developers of the computer industry at the end of the twentieth century. A major difference between integrated circuits and cellular design is that the fixes to errors in integrated circuit design and manufacture that could be (and usually were) made by firmware and software changes are not available to the cellular system designers. Simulations of the cellular design are now possible because of new high-speed parallel computers (gigaflop/microsecond), so the design and testing cycle can be closed in a relatively short period.

Computational power, which is now finally matched to the complexity of biological systems, has increased sufficiently to also allow simulations of integrated circuits for more than just a few microseconds so that circuit design errors can now largely be corrected before manufacture, and firmware and software fixes are less prevalent. Biology has entered a new mature phase because of the convergence of high-throughput data acquisition and computing technology. Engineering in biology has finally emerged as a sophisticated technical enterprise. The era when computing power finally reached the threshold necessary to compute biological systems of real complexity is now recognized as a major milestone in human scientific capability. This revolution has eclipsed in recognition and importance by orders of magnitude the sequencing of the human genome, completed in 2000.

Some of the things that are being done with engineered cells would have been unexpected in 2000. It was anticipated that the use of specifically engineered cells would find wide application in medicine to correct a defect or repair a damaged organ or joint. These applications have taken an unexpected turn. Medical scientists are now designing organs that were not produced by evolution and that do not exist in normal humans to correct a pathology, deflect a degeneration, or enhance a biochemical capacity of the body.

Other broader uses were not anticipated at all. For example, we have learned how to control the processes of biodirected crystallization that many organisms use to make exquisitely precise nanostructures of silicon, calcium carbonate, and composite materials of these with proteins (Cha et al., 2000). The genetically controlled architecture of the structures of the cocolithophoids of the oceans and marine diatoms and sponges are now controllable at the DNA level, and engineered organisms can produce precise and well-defined structures for use in a variety of micro- and nanotechnological, mechanical, and electronic applications.

Another application derives from the Turing machine-like design for a DNA computer proposed in 1999 (Liu et al., 2000). The device is now engineered into an artificial cell that signals the input and output to its internal DNA computer through the cell's surface receptors. In a very short time the advance of the life sciences has been transformed into advanced applications well beyond the medical or agricultural. This future realm of converged technologies and sciences has brought with it a powerful reminder of both the immense power of understanding and the responsibility for societal control of its applications.

BIOLOGY AS AN INFORMATION SCIENCE

The above extrapolation to a brave and very new world of the life sciences is based on only slightly more advanced technologies than we have today, albeit with some significant license and no assurances of accuracy. However, these extrapolated technologies, when combined with some fundamental insights of a new biology, can clearly make an enormous difference. The realm is actually very plausible and is one that may not be as far away as we might imagine. One basic idea about biology that underlies our thinking is that it is essentially an information science. This statement should not be taken in the trivial sense by which any science can be described as an information science in that descriptions of phenomena and prediction all involve handling information, but rather it should be considered as an assertion of the fundamentally different properties of biological systems. We would argue that it is fundamental to the way in which we should think about biological problems as distinct from, for example, physics, chemistry, geophysics, astronomy, and cosmology. A full discussion of this assertion is beyond the scope of this paper and will be addressed elsewhere, but because there are important lessons to be derived from the conclusions, the argument will be summarized here. The main arguments for the assertion that biology is an information science are as follows:

- All biological processes can be described as information transactions.

—Inheritance, recombination, transcription, translation, signaling, etc., all involve specific processing and transfer of encoded information.

• Biological entities are often temporally stable even though their physical components turn over. Organisms are transient and species are longer lived. Similarly, the molecular constituents are transient, yet the information they encode is long lived.

—The information encoded in the genomes of the members of a species population is the essential definition of that species, and the information changes significantly only as evolution produces new or different species.

—The fidelity of biological processes is linked to the lifetime of the information-carrying species. For example, genes are long lived, so replication must be extremely accurate. Messenger RNA carries information for briefer periods and in multiple copies, so transcription need not be quite as accurate. Because of protein turnover, translation is not as stringent in its proofreading capabilities.

• Evolution changes the information content of genomes.

—The ways in which variation and selection act to change the information in the genome are at the heart of the evolutionary process—the loss of information through random variation balanced against the addition of information by selection based on phenotype.

• Biological systems are characterized by an irreducible complexity.

—The encoded instructions for cellular life involve the direct and indirect interaction of a large number of molecular components behaving as a complex system with many emergent properties.

When presenting biology as an information science, one must ask: Is this just a new expression of old ideas or are there underlying principles to be asserted from this new perspective? It has been argued that biology is just an extension to a higher level of organization of physics and chemistry, albeit in systems far from equilibrium. So where does information science come in? Biological systems represent complex, nonlinear, and far-from-equilibrium systems. They also represent systems that are stably replicated over long stretches of time from encoded information.

Until recently, such systems have been deemed too "messy" for proper exploration by the mainstream of physics and chemistry. The key to putting order to the wealth of data in the biological sciences is through an understanding of the dynamics of complex systems. One of the crucial characteristics of species of living organisms is the ability to evolve in response to external environmental pressures. At the same time, a distinguishing feature of most organisms is their ability to buffer their internal components and structures against external fluctuations. Biological systems are adaptive yet stable. Any understanding of biology must, by necessity, deal with the stability conditions for complicated nonlinear systems. Currently, these problems are too complex for traditional physicochemical approaches.

To assess the stability of a dynamic system, one can use what are called Lyapounov functions. The Gibbs entropy is an example of a Lyapounov function used to describe the stability of thermodynamic systems. Recent work suggests that for some biological applications the Shannon information can also serve as a Lyapounov function (Lehman et al., 2000). This allows the stability of the evolution of the system to be described in a direct and simple manner. In such cases we do not need to return to thermodynamic arguments but can instead use approaches from information theory and the dynamics of complex systems. This provides a theoretical foundation for presenting biology as an information science. The analysis of such fundamental issues is, perhaps, the biggest challenge in developing a quantitative biology and in understanding what our models mean, what constraints are in effect in biological systems, and how best to shape theory to reflect experiment in biology.

These arguments may seem too theoretical to have practical consequences, but the consequences of the central role of information in the biological sciences are indeed practical and far reaching. They have significant implications for how we approach overcoming the many technical obstacles that remain to achieving the vision of the future of the life sciences described above. They have significant implications as well for the essential bridging of the disciplinary gaps between biology and the other sciences. For example, as an educational tool, the formulation of biology in these terms represents an exciting challenge that could help greatly in building robust permanent bridges between biology and the more mathematically based physical sciences and engineering. These gaps are significant impediments that need to be aggressively closed. Finally, the approach to biology as an information science should have important implications in harnessing the power of the biological sciences for practical applications.

THE DATA AVALANCHE

It has become commonplace to cite the tremendous increase in biological information available today and to point out the increasing rate of its acquisition. This accelerating avalanche is, in fact, a discipline-altering change that deserves acknowledgment. Yet it is also an enormous problem. At the moment, there are more 100 databases available on the World

Wide Web that hold useful information about the genetic contents and biochemical components and systems of various cell types. More than 25 full microbial genome sequences have been completed and are available in public databases, and probably more than 50 are currently in the process of being fully sequenced. Full genomic sequencing has become a powerful tool in approaching very practical environmental and therapeutic problems related to microorganisms. We expect that in the next few years at least a few hundred complete microbial genomic sequences will be available.

The Human Genome Project provides a powerful illustration of the rate of acceleration of information acquisition in the past few years. Before 1993 there were a few million base pairs of human sequence data in the databases (some of them redundant), and the 5-year goals for the project, as set out in 1993 (Collins and Galas, 1993), posted a rate of 50 million base pairs per year by the end of 1998. By then more than 100 million base pairs were entered into the database. The year 2000 is an historic one, as during this year the entire human genome sequence will be essentially (90 percent) complete.

In March 2000 the completed sequence of the *Drosophila* genome was published (Adams et al., 2000). In April three more human chromosome sequences were completed, bringing the completed score to four, and the remaining sequences are expected by year's end. The millennium turns with our having the full sequence of our own genome available for our scrutiny and understanding, a truly historic and transforming event for both the basic and applied sciences.

It is perfectly within reasonable limits of projection to suppose that hundreds of microorganism genomes will be fully sequenced each year in the first decade of the new millennium. Indeed, the capacities of the major sequencing centers are such that the equivalent of one full microbial genome project is completed each day. These kinds of data, genomic sequence, are, however, only the very simplest kind of data—digital, one-dimensional data. The much more complex data on protein interactions, gene expression levels, cell surface receptor populations, the spectrum of genetic variations in populations, and other subtle biological complexities are the all-important informational font for the biology of the future. This information will occupy the basic and applied life sciences for decades to come.

NEW TECHNOLOGIES: THE GENOTYPING EXAMPLE

As argued above, the life sciences are unique among the sciences in being driven almost entirely by technological change. The field of human genetics is one of the areas undergoing major shifts now as a result of the recent discoveries of the extent of single nucleotide polymorphisms (SNPs) in the human genome and the development of methods for identifying and scoring (genotyping) these genetic variations. It is useful to describe a specific example of the changes in the technology for genotyping in order to explore both the role of interdisciplinary science and the scale of change in data acquisition rates now upon us.

We cite here one example of genotyping technologies—the use of cleavable small molecule tags for DNA that have been developed for genotyping large numbers of samples automatically and simultaneously scoring variations at many sites per sample. This technology is one example of the rapid acceleration of throughput and automation that is being seen in many areas of biological technology. Others include automated enzyme-based assays with fluorescent readouts, different mass spectrometry methods, and others. The new technologies will be cheaper, faster, and more miniaturized than these.

The underlying idea of this mass-tagging technology is essentially that detecting and discriminating among small molecules is an easier problem than detecting and discriminating among large nucleic acid molecules such as DNA fragments. Thus, a tagging strategy is used in which a large number (hundreds) of small molecules are attached via a photocleavable linker to specific DNA oligonucleotides in a scheme that encodes the sequence of each of them. The assay that selects one or the other of the oligonucleotides corresponding to the genetic variant at each site (SNPs) is completed when the tags are cleaved from the DNA with intense light and detected in a mass spectrometer. The genotypes are then reconstructed with a simple computer algorithm, and the data are stored for analysis.

The technology has been adapted to a fully automated scheme that, in the first-generation system, can read more than 40,000 genotypes per day per machine. In addition, it is clear that this technology, like many others of a similar age, will be amenable to substantial changes in efficiency, speed, and cost in the near future. This kind of automated parallelized data gathering is a good quantitative indicator of things to come in this area. Large amounts of genotype data for human populations at loci that are well characterized are already being used to reconstruct human prehistoric population migrations, to track down genetic disease loci and loci that determine susceptibilities to a variety of conditions and therapies. The realization that it may be possible to find genetic loci and specify variations at certain of these loci that determine differential response to therapeutic drugs has triggered an extensive exploration of the field now dubbed "pharmacogenomics." While it is very early in this historic endeavor, it is already clear that medical treatments in the future will have to deal with the genetic variations among patients to maximize therapeutic effectiveness and avoid patient-specific toxicity and lack of efficacy. Health care in the future will be dependent in many areas on this type of technology for essential patient information.

KEY ROLE OF INTERDISCIPLINARY EDUCATION AND TRAINING

It is clear from the above discussion that if our view of the future is at all accurate we will need an abundance of people trained in the integration of several disciplines who have adopted the data, technology, and modeling philosophies for the life sciences described here. Without effective integration of these disciplinary elements and the ability to function well outside the usual boundaries of the biological sciences, this type of training will fail.

A new academic institution is being developed specifically to provide this training. It is the goal of the newest Claremont College, the Keck Graduate Institute of Applied Life Sciences (KGI), to play a key leadership role in the development of the new area of applied biology and to train professional leaders in this area. This goal is distinct from that of training the next generation of researchers, but it has some features in common with that important goal. The professional training provided by KGI's two-year master's of biosciences degree is directed toward broad-based technical training, integrated with management and ethics—an objective that is increasingly recognized as key (Rayl, 2000).

The technical training must be fundamentally integrated, not just a collection of conventional discipline-specific courses, and intensive enough to enable an understanding of the key science and many of the important questions posed by real problems. As planned, the program will not attempt to train researchers, in either the depth required or the orientation, but the integrity of the advanced scientific training required must be paramount.

The effective training of a new breed of professionals in the applied life sciences is an important way to attempt to fulfill this leadership role, since people are the key to this new era. As such it will be important for the KGI to nurture close ties to various industrial, academic, and government institutions to track the burgeoning development of applied biology and to collaborate with researchers and managers to understand the changes afoot in the applied arena. These professionals are intended to play a role in the applied life sciences similar to that played by the master's engineers in the physical sciences-based industry. The KGI training program will require significant changes in the ways in which training is conducted. Some features of these changes will include problem-centered training—that is, an educational experience that has as its unifying theme the solution of practical real-world technical problems requiring the application of diverse disciplinary elements, close team work, and concentrated effort for a limited time period. The training will attempt to illustrate by example such modes of working. Both academic and industrial involvement in this mode of training will clearly be essential to its success. Industry internships and collaborations with both industry and academic colleagues will provide important occasions for this training. In many ways the model of engineering training is most illustrative of this training philosophy, but the information sciences are perhaps the closest to our intellectual philosophy.

Cooperative interdisciplinary problem solving will necessarily introduce the important topic of the behavior of organizations, as well as project and people management. In addition, some of the significant social issues related to applied biology, particularly in areas such as human genetics, health care resource issues, clinical trials, and environmental impacts, will arise naturally in these endeavors. All of the abovementioned areas will be treated more fully by some focused program elements; by project-based training; by using affiliated institutions, adjunct faculty, and external industry associates; and by direct industrial experience for students.

The intellectual foundations of the KGI, the axis around which we can orient our development in applied biology, derive directly from new developments in fundamental biology to which a wide variety of other disciplines contribute. The challenges of this new institution, representative of all who dedicate themselves to these goals, are to focus on applications and innovative professional training while tracking new developments in the life sciences in all sectors and establishing a strong interdisciplinary applied research program to complement and invigorate the training program. It is a challenge matched to the changes taking place in the biological sciences.

CONCLUSIONS

The hallmarks of the future of the applied life sciences are high-throughput technology for data acquisition, massive data acquisition and storage, and the use of multiple technical disciplines in both the development of these technologies and the analysis and application of the data generated. Advances in automation and miniaturization and the increasingly interdisciplinary nature of technological developments are key to the future of the life sciences, largely because the inherent complexity of biological systems requires a massive increase in an already high rate of data acquisition. Analysis of these data and their synthesis into new scientific knowledge will dominate the foreseeable future of the basic biological sciences.

An essential scientific and technological activity is the development of a modeling capability for complex biological systems that greatly exceeds anything we now have. The move toward integration in all the above senses is a key to the future. We have ventured to speculate about what the future holds in these areas of biological sciences and their applications. These speculations may be flawed in detail, but we suggest that some aspects are almost certain to be realized. They are intended to serve to stimulate current thinking about this future, its implications, and the needs it implies.

We anticipate that the near future will bring major shifts in how the life sciences are conducted in both their basic and applied forms. The applications that are at center stage in this volume—those focused on the understanding and control of infectious diseases and related issues in human health, environment, and agriculture—are specific instances of a much broader area of application of these ideas. People are key to the future. Cross-disciplinary education and training of a new breed of professionals in the applied life sciences are crucial to achieving this future. Programs such as the one at the KGI Institute of Applied Life Sciences have taken on this challenge, one that will be necessary to meet in order to realize the promise of the convergence discussed here.

REFERENCES

Adams, M. D., S. E. Celniker, R. A. Holt, et al. 2000. The genome sequence of *Drosophila melanogaster*. Science, 287:2185-2195.

Arnheim, N., and H. Erlich. 1992. Polymerase chain reaction strategy. Annual Review of Biochemistry, 61:131-156.

Beaudry, D. P., and G. F. Joyce. 1992. Directed evolution of an RNA enzyme. Science, 257:635-641.

Boyer, H. W. 1971. DNA restriction and modification mechanisms in bacteria. Annual Review of Microbiology, 25:153-176.

Cairns, J., G. Stent, and J. D. Watson, eds. 1966. Phage and the Origins of Molecular Biology. New York: Cold Spring Harbor Laboratory.

Cha, J. N., G. D. Stucky, D. E. Morse, and T. J. Deming. 2000. Biomimetic synthesis of ordered silica structures mediated by block copolypeptides. Nature, 403(6767):289-292.

Cohen, S. N., A. C. Y. Chang, and H. W. Boyer. 1973. Construction of biologically functional bacterial plasmids in vitro. Proceedings of the National Academy of Sciences USA, 72:3240-3245.

Collins, F., and D. J. Galas. 1993. A new five-year plan for the U.S. Human Genome Project. Science, 262(5130):43-46.

Itakura, K., J. Rossi, and R. B. Wallace. 1984. Synthesis and use of synthetic oligonucleotides. Annual Review of Biochemistry, 53:323-356.

Köhler, G., and C. Milstein. 1975. Continuous cultures of fused cells secreting antibody of predefined specificity. Nature, 256(5517):495-497.

Lehman, N., M. D. Donne, M. West, and T. G. Dewey. 2000. The genotypic landscape during in vitro evolution of a catalytic RNA: Implications for phenotypic buffering. Journal of Molecular Evolution, 50:481-490.
Liu, Q., L. Wang, A. G. Frutos, A. E. Condon, R. M. Corn, and L. M. Smith. 2000. DNA computing on surfaces. Nature, 403(6766):175-179.
Maxam, A. M., and W. Gilbert. 1977. A new method for sequencing DNA. Proceedings of the National Academy of Sciences USA, 74:560-564.
Nathans, D., and H. O. Smith. 1975. Restriction endonucleases in the analysis and restructuring of DNA molecules. Annual Review of Biochemistry, 44:273-293.
Rayl, A. J. S. 2000. From implants to explants, and beyond: Multidisciplinary panel emphasizes follow-up research. The Scientist, 14(5):16.
Sanger, F., and A. R. Coulson. 1975. A rapid method for determining sequences in DNA by primed synthesis with DNA polymerase. Journal of Molecular Biology, 94:441-449.
Southern, E. M. 1974. An improved method for transferring nucleotides from electrophoresis strips to thin layers of ion-exchange cellulose. Analytical Biochemistry, 62(1):317-318.
Watson, J. D. 1968. The Double Helix. London: Weidenfeld and Nicolsen.

21

New Standards and Approaches for Integrating Instruments into Laboratory Automation Systems[1]

Torsten A. Staab and Gary W. Kramer

INTRODUCTION

Plug-and-play laboratory instruments are the keys to reducing integration efforts and to increasing connectivity, reuse, and scalability. This approach requires instrument manufacturers to agree on standards for controlling laboratory equipment. They also need to agree on a standardized way to describe and exchange system and device capability information. There are two approaches to interfacing standards: prescriptive and descriptive. The former approach seeks to make all devices look and behave in the same fashion to the controller. The latter concept is to employ a common way to describe the idiosyncrasies of a device and its behavior. Because it is not possible, or even desirable, to make every automation device look or behave the same, other means must be found to accommodate differences.

---

[1]*Los Alamos National Laboratory (LANL) Disclaimer of Endorsement:* Reference herein to any specific commercial products, process, or service by trade name, trademark, manufacturer, or otherwise does not necessarily constitute or imply its endorsement, recommendation, or favoring by the U.S. government or the University of California. The views and opinions of authors expressed herein do not necessarily state or reflect those of the U.S. government or the University of California and shall not be used for advertising or product endorsement purposes.

National Institute of Standards and Technology (NIST) Commercial Disclaimer: Certain commercial equipment, instruments, and materials are identified to describe adequately the work presented herein. Such identification does not imply recommendation or endorsement by NIST, nor does it imply that the equipment, instruments, or materials are necessarily the best available for the purpose.

This paper describes three standardization efforts that address common instrument interfacing and integration problems. The first standardization effort discussed herein is the ASTM (American Society for Testing and Materials) E1989-98 standard, also known as LECIS (Laboratory Equipment Control Interface Specification; ASTM, 1999). LECIS defines a uniform deterministic remote control protocol for laboratory equipment. The other two standardization efforts, which are headed by the National Institute of Science and Technology (NIST), are complimentary to the ASTM LECIS standard. These efforts are known as the Device Capability Dataset (DCD; Staab and Kramer, 1998a) and the System Capability Dataset (SCD; (Piotrowski et al., 1998). The DCD defines a meta-dataset structure that captures device-specific information, such as physical dimensions, command sets, events, and resources. The SCD, on the other hand, describes the system-level capabilities. In general, these are the capabilities that are obtained by combining DCD information.

ASTM E1989-98 LECIS STANDARD

The dream of every laboratory automation engineer is to be able to integrate laboratory equipment with the plug-and-play ease of most of today's computer printers. Toward this end, the ASTM E1989-98 LECIS was developed to standardize equipment control. LECIS was developed as a generic standard that is applicable to a wide variety of equipment, from simple electronic balances to complex analytical instruments. The LECIS-defined instrument control protocol standardizes the application layer (see Figure 21.1), which is equivalent to the application layer in the OSI (Open Systems Interconnection) reference model. This standard has the advantage of being independent of low-level message-passing communications protocols. Typically, instrument message passing is accomplished by protocols such as RS-232 or TCP/IP. In the future, instruments may also be equipped with a USB (Universal Serial Bus) or an IEEE 1394 (Firewire) interface. The application layer approach has the advantage that it does not matter whether instruments and controllers are connected via a network, RS-232, or any other low-level interfacing standard.

ORIGINS OF ASTM E1989-98 LECIS

The LECIS was developed as a result of years of work by the NIST Consortium on Automated Analytical Laboratory Systems (CAALS; Kingston, 1989; Salit et al., 1994; Salit and Griesmeyer, 1997). One of the products of this industry consortium was the definition of a standard Common Command Set (Staab et al., 2000). This recommended command set was formalized into the LECIS standard by the ASTM E-49-52 Sub-

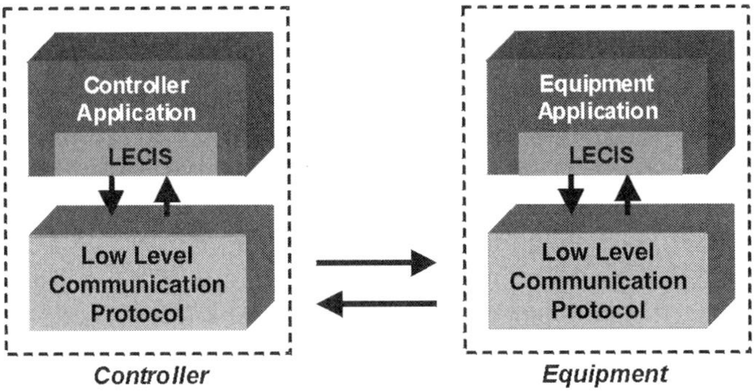

FIGURE 21.1 Protocol hierarchy.

committee on Computerization of Analytical Sciences Data of the ASTM E-49 Committee on Computerization of Material and Chemical Property Data (ASTM, 1999). The LECIS defines the standard equipment behavior and the messages that coordinate them in terms of the interaction model described in part by the General Equipment Interface Specification (GEIS; Griesmeyer, 1997) and in part by CAALS (Staab and Kramer, 2000). The GEIS was developed by the Intelligent Systems and Robotics Center at Sandia National Laboratory.

Control Paradigm

The LECIS instrument control concept is based on the following interactions:

- States (discrete operational equipment states).
- Commands (controller messages).
- Events (equipment messages).

The standard LECIS interactions are described in the form of state diagrams, which standardize the sequence of equipment behaviors, in terms of states, and the message passing that synchronizes the execution of these behaviors, in terms of transitions between states. Figure 21.2 shows the basic elements of an interaction state diagram. ASTM E1989-98 LECIS standard-compliant instruments must support every interaction defined by the standard. The interaction models provide defined behaviors by "falling through" states that are not appropriate for individual

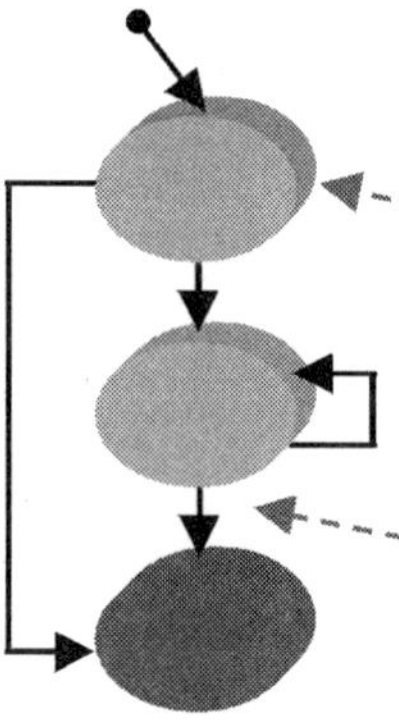

FIGURE 21.2 Interaction state diagram example.

equipment. This requirement allows the instrument controller to follow a standard control model regardless of equipment type.

Interaction Categories

The ASTM E1989-98 LECIS standard specifies required interactions only. Required interactions are interactions that all instruments must support. Equipment manufacturers can also use the LECIS interaction model to delineate equipment-specific behaviors. Equipment-specific interactions are optional interactions (see Figure 21.3).

Primary Interactions

The LECIS standard differentiates between primary and secondary interactions. Primary interactions are interactions that are started when an instrument is turned on and do not terminate until the instrument is powered down. Only one instance of a primary interaction can exist at any time. An analogy to a primary interaction is the old single-user MS-DOS operating system. When a personal computer is booted, the MS-DOS command interpreter is loaded and invoked and remains running until the PC is turned off. There can be only one command interpreter running at any time.

The ASTM E1989-98 standard defines the following two primary interactions:

- *Control flow* (regulates instrument initialization, configuration, and regular operations; see Figure 21.4).

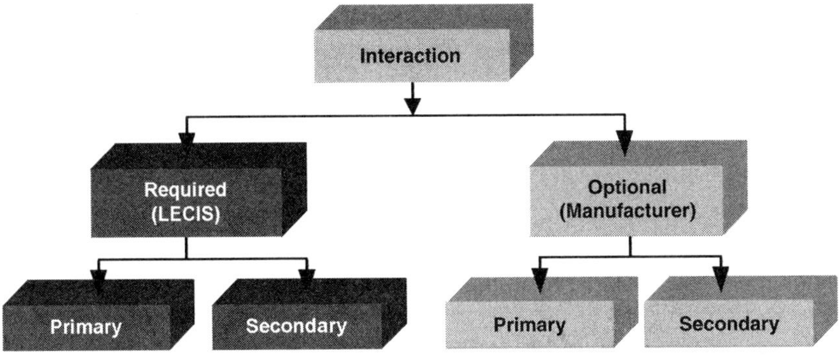

FIGURE 21.3 Interaction categories.

- *Local/remote control mode* (manages instrument's local/remote control mode transitions).

The primary control flow interaction, depicted in Figure 21.4, has three levels. The top level represents the regular instrument control flow. For example, after the instrument has been powered up and its communications link to the controller established, the instrument will be in a state called "powered up." After receiving the standard INIT command, the instrument transitions from the "powered-up" state to the "initializing" state. After completion of the initialization behavior, the instrument sends a standardized event notification to the controller and transitions to the "idle" state. In the "idle" state the instrument waits for the SETUP command to start the configuration process. With the remaining standard messages, the instrument proceeds through "configuring" to "normal OPS" (OPS = Operational State), where it remains until commanded to clear itself and reenter "idle."

States at the intermediate level of the control flow interaction define standard behaviors to pause (i.e., temporarily hold operations) and resume an instrument. If the controller commands the instrument to pause, the instrument remains in the current state in the central level while entering the "pausing" state.

The outermost level of the primary control flow interaction, shown in Figure 21.4, contains only the "Estopped" state. The "Estopped" state is used to represent an emergency shutdown. Note that a transition to "Estopped" is unidirectional, in accordance with standard safety requirements. Once in "Estopped" the local instrument operator must intervene to restart the instrument so that it reenters the "powered-up" state.

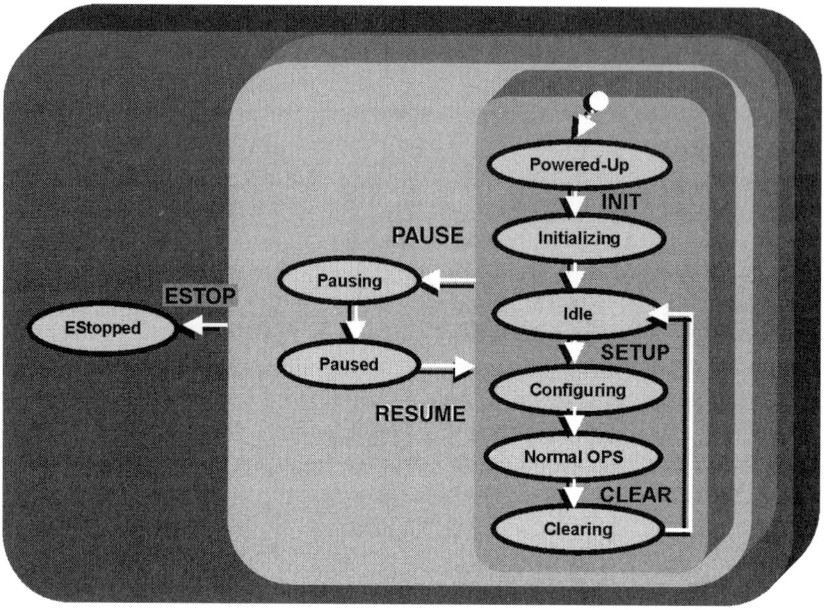

FIGURE 21.4 Control flow interaction.

Secondary Interactions

Secondary interactions are interactions that can be started and terminated at runtime. It is also possible to create multiple instances of the same interaction, allowing standard control of multitasking behavior. The ASTM E1989-98 standard defines the following required secondary interactions:

- Processing (allows execution of instrument methods).
- Status (retrieves status of interactions from the instrument).
- Lock/Unlock (locks/unlocks instrument's data/material ports).
- Item Available Notification (data/material availability notification).
- Abort (aborts interactions).
- Alarm (indicates instrument errors/exceptions).
- Next Event Request (requests the next event from instrument).

ASTM E1989-98 LECIS Status and Directions

As a proof of concept, LANL successfully implemented the ASTM E1989-98 LECIS standard on an internally developed universal microtiter plate handling device.

The NCCLS (formerly the National Committee for Clinical Laboratory Standards) Area Committee on Laboratory Automation has incorporated the LECIS behavior definitions in its proposed AUTO3-P Communications with Automated Clinical Laboratory Systems, Instruments, Devices, and Information Systems standard.

LANL is currently working, with varying degrees of formality, with Bristol-Myers Squibb, CRS Robotics, Inc., ErgoTech Systems, Inc., and Pfizer to facilitate commercial adoption of the LECIS standard. Bristol-Myers Squibb intends to build standard-compliant, Active-X-based controls for its laboratory instruments. CRS Robotics, Inc., is committed to adding an ASTM E1989-98 interface to POLARA laboratory automation software. ErgoTech Systems, Inc., is developing a supported open source reference software implementation of the standard. Pfizer is currently working on a TCP/IP-based LECIS reference implementation.

The goal of this collaborative effort is also to develop CORBA (Common Request Broker Architecture) and DCOM (Distributed Component Object Model) IDL (Interface Definition Language) versions of the ASTM E1989-98 standard. The CORBA and DCOM IDL standard extensions will further simplify the development and implementation of new laboratory equipment driver software.

Many companies have expressed interest in the new ASTM E1989-98 instrument control standard. More information and progress reports on the development and implementation of the standard, as well as related projects, are available on the Web at *www.lecis.org*.

THE DCD

A DCD describes the idiosyncratic characteristics of laboratory equipment, such as the equipment's identity, physical dimensions, location, supported command set, generated events, input/output (I/O) ports, and other resources. The DCD concept provides a means for standardizing the interfacing of laboratory automation devices in a descriptive rather than prescriptive manner.

DCD Origins

The first steps toward the definition of a standard DCD were taken by participants of the Consortium on Automated Analytical Laboratory Systems Modularity Working Group at NIST in 1994. Since then, research on the DCD topic has continued at NIST and the Fachhochschule Wiesbaden (FHW) University of Applied Sciences (Giegel, 1996; Hagemeier, 1998; Schäfer, 1996). Initial results of this joint research effort were presented at the LabAutomation'98 conference in San Diego in January 1998. The SCD, a further extension of the DCD idea that accommodates individual DCDs as well as interactions, roles, and dependencies between devices in a system, is described later in this paper. Figure 21.5 depicts the generic CAALS logical control architecture (Salit et al., 1994) showing the relationship of the DCDs and the SCD.

Structure of DCDs

A DCD, comprised of various types of information about the device, its operations, and its resources, can be organized into the following categories:

- Identification
- Communications

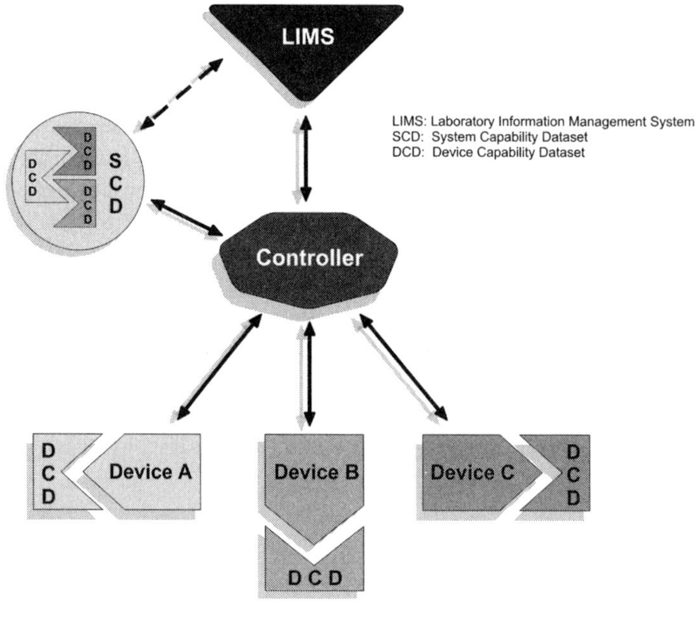

FIGURE 21.5 CAALS logic control architecture.

- Physical characteristics
- I/O ports
- Commands
- Events
- Exceptions
- Errors
- Resources
- Maintenance

In programming language terminology, such categories are the equivalent of classes or structures. Representative data and their data types contained in these DCD categories are detailed elsewhere (Staab and Kramer, 1998a, 1998b). The above DCD categories are still under development and therefore are neither final nor complete.

Static Versus Nonstatic Device Capability Data

Not all the data in DCD categories have the same temporal nature. Some information, such as the device manufacturer's name, will not change throughout the lifetime of the device and so can be considered static. Other facts, such as resource amounts and consumables, change during the operation of the device and must be accounted for dynamically. There are cases between these two extremes that range from semistatic to semidynamic data. Analytical method names could be an example of semistatic data, while a device's initial fill level might constitute semidynamic data. The exact temporal attributes assigned are less important than the concept that data in a DCD can vary in both time of inclusion and longevity. A multiauthored, multisession creation DCD process is implicit in this notion. A DCD is created along a multistep path beginning with the instrument manufacturer, who imparts much of the static data. The system integrator and methods developer supply additional static, semistatic, and semidynamic data. And the system itself, during run time, updates many of the dynamic data.

DCD Representation and Access

Currently, no consensus exists regarding the format for the storage and representation of DCDs. Whichever format is ultimately chosen for the DCD representation, the standard should be flexible, extensible, machine interpretable, and computing platform independent. To make the DCD idea work, we will either have to standardize on a single abstract metadata representation scheme or utilize compatible schemes that can be reliably interconverted electronically. (Metadata are data about data;

for example, the metadata description of a field in your address book would include the maximum length of the field, acceptable fonts to be used, etc.). Data and interface modeling languages such as the Extensible Markup Language (XML; *www.w3.org/XML*), the IDL (*www.omg.org*), or EXPRESS (ISO/TR, 1997; Schenck and Wilson, 1994; ISO 1994a, 1994b) are all current contenders for the abstract representation of a DCD.

DCD Access and Location

DCDs can be stored in anything from a simple ASCII file to a high-performance relational or object-oriented database. An entire DCD need not be constrained to a single type of storage representation—the static and dynamic DCD information could reside in different locations on different media. For example, the static parts of the DCD information might be stored in an instrument's ROM (read-only memory), on disk, or even on a networked server provided by the device manufacturer. The dynamic part could reside either in flash memory, on disc, or in random access memory.

It is crucial for the controller to have access at run time to the DCDs of all devices under its control. A controller could access a device's DCD either directly using special data commands or indirectly through normal control channel communications.

Dynamic DCD information is more likely to reside in the instrument's local controller. The system-level device controller could retrieve such information either on demand or as a result of receiving an event message. On demand means that the controller explicitly asks the device for DCD information, and this request initiates a download of the static and dynamic DCD information from the instrument. In the event-driven model the device informs the controller each time a DCD characteristic changes by sending an event message.

DCD Status and Directions

NIST recently published its initial DCD specification (Staab and Kramer, 1998b). The Fachhochschule Wiesbaden in Germany and NIST are currently working on the first CORBA IDL-based DCD prototype implementations. Pharmaceutical companies, such as Pfizer, are working on XML-based implementations of the DCD concept. The trend toward plug-and-play instruments and component-oriented software will further foster the DCD adoption process.

THE SCD

At the system level, a controller needs to know the characteristics of the devices that it controls. However, providing the controller with access to the DCD of each instrument is not sufficient. Another construct that accounts for interdevice relationships, dependencies, and resources that are shared among devices but not part of any particular device is needed. This construct is called the SCD (Richter et al., 2000a, 2000b).

A system is more than just the sum of its parts. The system controller must have a means to account for the independent devices or SLMs (Standard Laboratory Modules; Salit et al., 1994), support devices or SSMs (System Support Modules, such as a robot or system-wide machine vision system), and system-level resources such as samples, common reagents, and storage areas (Schäfer et al., 1998). To be robust, the system controller must provide sample and resource tracking, deal with result data, and handle errors. The system controller needs a mechanism to provide its capabilities to the LIMS or the next higher level of control. Extending the DCD concept to the system level can provide all these features. As with a DCD, the SCD is comprised of building blocks, as shown in Figure 21.6. The capability dataset "puzzle" shown there is not complete—the missing piece(s) represent our current lack of understanding of what will ultimately be required to adequately represent the device or system being described in the capability dataset.

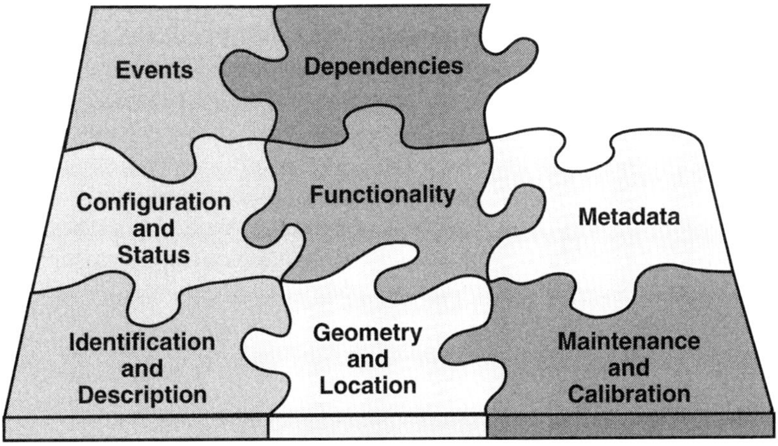

FIGURE 21.6 Components of a capability dataset.

SCD Components

Identification and Description

The identification components are used for configuration management and audit information for devices in the laboratory automation system, including the manufacturer's name, device name, device software version, and so forth.

Geometry and Location

This block holds the descriptions of physical size, shape, and location of objects and their accessible parts. Robotic paths or curves for accessing the device's ports are also parts of this block. At the system level this information is used for robot collision avoidance.

Maintenance and Calibration

Information about maintenance and calibration is used for validation and quality assurance purposes.

Configuration and Status

Both hardware and software configurations are stored in a capability dataset. The capability dataset contains configuration constraints as well as the current configuration of the objects. These data ease the configuration of a device and reduce the chances for incorrectly configuring a device. Configuration parameters are needed to set the proper functionality of a device via user methods. Status information is usually used to monitor the actions of the devices on the workbench.

Functionality

This block describes commands and their functions along with the types and ranges of any associated command data. This information enables the controller software to interact with the device because the software can both know the commands and how to use them.

Metadata

Capability datasets contain metadata for any data that can be sent from a device and that must be processed by the system. These metadata include results data as well as status data, maintenance data, and calibration data. Metadata describe both the type and the structure of data.

Events

Device events are messages sent from the device to the controller. They are categorized into different classes such as warning, alarm, error, and data.

Dependencies

This block represents concepts such as ownership management—who can access which resource and when. Also, dependencies between different devices are stored here. For example, if the system treats a chromatograph and its autosampler as two devices, the chromatograph is dependent on the autosampler for its operation and may become useless if the autosampler fails, even though the chromatograph itself is still operable.

Static and Dynamic SCD Contents

Information enters the SCD at differing times. Some of the data are available at the time the system is designed and never change. This static part includes capabilities of the system's devices such as command sets, events, and geometry. Other SCD information is dynamic and is not known until run time—for example, sample locations, actual status of instruments, and current amounts of reagents on the workbench. Device configurations can change constantly, and even the current position of some components on the workbench can change. It is essential that the SCD provide a means for keeping track of components and their positions on the workbench. The SCD must track which device has permission to use a shared resource or port.

SCD Creation

Figure 21.7 shows a time line for the creation of an SCD. First, the device manufacturers provide the information that describes the identification, functionalities, and general behavior of the devices. This information forms the core of the DCDs for the devices. Next, the system integrator combines all of the DCDs and adds system-specific information about each device's role, position in the system, error handling, error mapping, and ownership of resources. "Superusers," who develop analytical methods, set up the system with analytical procedures and add the information about them. Thus far, most of the information incorporated into the SCD

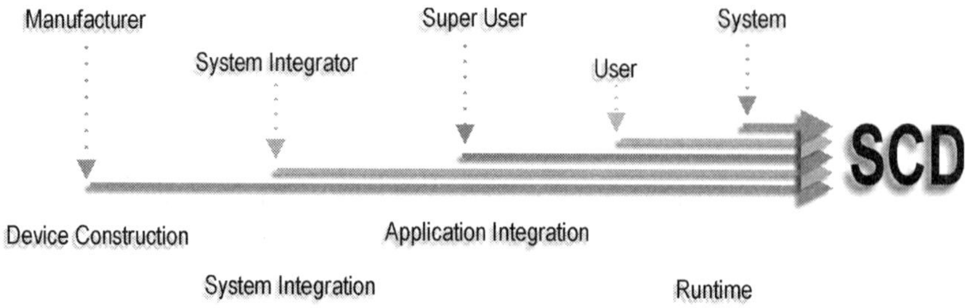

FIGURE 21.7 SCD creation timeline.

is static. Application users enter the parameters for specific runs, such as the number of samples or a specific temperature for a device. This sort of information is considered semistatic because although it is known when the run starts, the values can change as the run progresses. Finally, during run time the system supplies information to the SCD about its current state. These dynamic characteristics are primarily used for process monitoring.

SCD Representation

To construct a capability dataset that is easily reusable from system to system without significant reconstruction, device characteristics must be described as abstractly as possible. The representational scheme should be independent of the device being described, and the device/system description must be completely independent from the implementation or the platform used. Once created in this way, device information can be used as long as the device itself is used. Even when the implementation changes, the device description is not lost. To achieve the implementation independence of the capability dataset, a neutral mechanism that is capable of describing product data is needed. The information modeling language EXPRESS, contained in ISO 10313 (STEP), fulfills this requirement (ISO/TR, 1997). It is an international standard for computer-interpretable representation and exchange of product data. It is very important to note that STEP focuses not only on what the data are but also on what they mean and how they relate to one another. Originally, STEP grew out of a need to exchange product data among different computer-aided design systems. However, now it is being used in a variety of applications from metal parts fabrication to chemical plant design. The wide scope of STEP applications means that many elements of STEP have already been invented by others. It is possible to utilize these elements, such as two- and three-dimensional geometric description schemes and tolerancing mechanisms that have already been developed and tested, without having to reinvent them. Conversely, elements specifically created for the SCD and DCD concept may be useful to others who want to utilize device characterizations and, perhaps more importantly, the descriptions of result data. This reusability of models is a big advantage to this approach.

SCD Implementation

The capability dataset approach purposefully isolates the description of the device and system from the implementation. EXPRESS, for example, provides a stable international standard for representing the SCD without worry about databases, types of databases (relational, object-oriented), or other implementation tools (ISO/TR, 1997). Furthermore, the SCD should be transparently accessible via a network in a distributed laboratory computing environment. This can be done by creating a standard interface for the SCD. NIST's current SCD approach defines the interfaces in CORBA-IDL (*www.omg.org*). CORBA provides the desired vendor, operating system, and programming language independence and functionality, and it is widely used in industry for the implementation of distributed systems. However, CORBA is not the only way to realize this interface. Other distributed component technologies such as Microsoft's DCOM also could be used. DCOM and CORBA are closely related system architectures and allow communication over software bridges. Since DCOM clients can easily access CORBA servers and CORBA is an independent standard, the interface definition is preferably done in CORBA-IDL.

SCD Status and Directions

The abstract EXPRESS data definitions of the SCD should eventually become part of an international standard, perhaps STEP, and should be flexible enough to allow a device manufacturer to describe all device properties and behaviors. An instrument manufacturer provides descriptions of the devices and components in STEP-like files using the SCD EXPRESS data definitions. This information is converted into transfer structures that can be understood by the capability dataset interface. These transfer structures are then stored in the underlying database in an implementation-specific format.

NIST's initial version of the SCD data definitions is complete (Staab and Kramer, 1998b). The next step, to implement the SCD concept for real devices, is currently under way at NIST in collaboration with the Wiesbaden

Computer Integrated Laboratory (WICIL) at the Fachhochschule Wiesbaden University of Applied Science (Tauber et al., 2000; Borchert et al., 2000).

CONCLUSION

In general, laboratory automation system integration is getting easier. The technologies presented here represent crucial building block technologies for the uniform, standardized, device-independent control of laboratory equipment. ASTM E1989-98 LECIS, DCD, and SCD are designed to foster the plug-and-play of laboratory instruments. Automated systems created by integrating equipment that follows these control and behavior models can be controlled more robustly and scheduled more effectively, increasing the reliability, usability, and throughput of automated systems.

Implementation projects that focus on these new technologies are currently under way at LANL, NIST, CRS Robotics, ErgoTech Systems, Pfizer, BMS, and FHW; elsewhere others are now being planned. The main purpose of these projects is to demonstrate the feasibility and real-world benefits of the ASTM LECIS, the DCD, and the SCD.

ACKNOWLEDGMENTS

The authors would like to acknowledge the work of the participants of the CAALS modularity workshops, who established the basis for standardization efforts—in particular, James R. DeVoe, Lawrence G. Falkenau, J. Michael Griesmeyer, Franklin R. Guenther, David R. Hawk, Richard S. Lysakowski, Marc L. Salit, Brian D. Seiler, William J. Sonnefeld, and John Upham. Furthermore, we are grateful for the valuable contributions of Anthony J. Day, Uwe Bernhöft, Peter J. Grandsard, Christian Piotrowski, Thorsten Richter, Thomas Tauber, Oliver Borchert, Mark F. Russo, John W. Elling, Tony J. Beugelsdijk, Robert M. Hollen, and Reinhold Schäfer.

This project is funded (in part) by NIST's Systems Integration for Manufacturing Applications (SIMA) Program. Initiated in 1994 under the federal government's high-performance computing and communications effort, SIMA addresses manufacturing systems integration problems through applications of information technologies and development of standards-based solutions. With technical activities in all of NIST's laboratories covering a broad spectrum of engineering and manufacturing domains, SIMA is making information interpretable among systems and people within and across networked enterprises.

REFERENCES

American Society for Testing and Materials. 1999. Standard Specification for Laboratory Equipment Control Interface (LECIS), ASTM E1989-98, ASTM Annual Book of ASTM Standards, 14.01, 915-939.

Borchert, O., R. Schäfer, and G. W. Kramer. 2000. Interface Definition of the System Capability Dataset Data Stream Schema. NIST Interagency Report. Washington, D.C.: U.S. Department of Commerce.

Giegel, B. 1996. Entwurf und Implementierung verteilter, objektorientierter Laborgeräteschnittstellen auf der Basis des CORBA-Standards. Diploma thesis, Computer Science Department, Fachhochschule Wiesbaden, Weisbaden Germany (*wwwvs.informatik.fh-wiesbaden.de/publications.html*).

Griesmeyer, J. M. 1997. General Equipment Interface Definition. Appendix A in Final Report: An Enabling Architecture for Information Driven Manufacturing. SAND97-2067, August 1997 (infoserve.sandia.gov/sand_doc/1997/972076.pdf).

Hagemeier, M. K. 1998. Entwurf und Implementierung einer generischen, objektorientierten Ansteuerung von Analyseautomaten unter Verwendung des CORBA-Standards. Diploma thesis, Computer Science Department, Fachhochschule Wiesbaden, Weisbaden Germany.

ISO. 1994a. Industrial automation systems and integration—Product data representation and exchange—STEP Part 1: Overview and fundamental principles, pp. 10303-10317.

ISO. 1994b. Industrial automation systems and integration—Product data representation and exchange—STEP Part 21: Implementation methods: Clear text encoding of the exchange structure, pp.10303-12157.

ISO/TR. 1997. Industrial automation systems and integration—Product data representation and exchange—STEP Part 12: Description methods: The EXPRESS-I language reference manual, pp. 10303-10312.

Kingston, H. M. 1989. Consortium on automated analytical laboratory systems. Analytical Chemistry, 61:1381A-1384A.

Piotrowski, C., T. Richter, R. Schäfer, and G. W. Kramer. 1998. The system capability dataset for laboratory automation system integration. Journal of the Association for Laboratory Automation, 3:51-55.

Richter, T., C. Piotrowski, R. Schäfer, and G. W. Kramer. 2000a. Interface Definition of Capability Datasets, Version 1.44. NIST Interagency Report. Washington, D.C.: U.S. Department of Commerce.

Richter, T., C. Piotrowski, R. Schäfer, and G. W. Kramer. 2000b. Implementation Guide to the System Capability Dataset, Version 1.15. NIST Interagency Report. Washington, D.C.: U.S. Department of Commerce.

Salit, M. L., F. R. Guenther, G. W. Kramer, and J. M. Griesmeyer. 1994. Integrating automated systems with modular architecture. Analytical Chemistry, 66:361A-367A.

Salit, M. L., and J. M. Griesmeyer. 1997. System ready behaviors for integration. Laboratory Robotics and Automation, 9:113-118.

Schäfer, R. 1996. Wiesbadener Computerintegriertes Labor (WICIL)—Entwicklung von Informatikmethoden zur Automatisierung Chemischer Analysen, Fachhochschule Wiesbaden, Germany, Ergebnisse aus Forschung und Entwicklung 27, ISBN 3-923068-27-1 (*www.informatik.fh-wiesbaden.de/labors/wicil.html*).

Schäfer, R., T. J. Beugelsdijk, and J. W. Elling. 1998. Architectural issues for total lab automation. Presented at the International Laboratory Automation 98 Conference, San Diego, Calif.

Schenck, D. A., and P. R. Wilson. 1994. Information Modeling: The EXPRESS Way. New York: Oxford University Press.

Staab, T. A., and G. W. Kramer. 1998a. The device capability dataset: A descriptive approach to laboratory automation system integration standards. Journal of the Association for Laboratory Automation, 3:45-50.

Staab, T. A., and G. W. Kramer. 1998b. Initial CAALS Device Capability Dataset, Version 1.07. NIST Interagency Report 6294 (*www.lecis.org*).

Staab, T. A., P. J. Grandsard, and G. W. Kramer. 2000. CAALS Common Command Set. NIST Interagency Report. Washington, D.C.: U.S. Department of Commerce.

Staab, T. A., and G. W. Kramer. 2000. CAALS High-Level Communication Protocol. NIST Interagency Report. Washington, D.C.: U.S. Department of Commerce.

Tauber, T., R. Schäfer, and G. W. Kramer. 2000. Device Communication Principles for Automation Systems Using a System Capability Dataset. NIST Interagency Report. Washington, D.C.: U.S. Department of Commerce.

22

High-Throughput Sequencing, Information Generation, and the Future of Biology

J. Craig Venter

When my laboratory at the National Institutes of Health started the first automated sequencing for the Human Genome Project, we had six 373 automated DNA sequencers that could run 16 lanes at a time, each once a day. Even at this rate, the biggest challenge for the project was interpreting the data and in the process finding the genes. In 1991 the expressed sequence tag (EST) method was developed, facilitating scale-up by using cDNA libraries to find large numbers of genes.

At the start of the 1990s there were fewer than 2,000 human genes known. Now there are millions of ESTs in the databases. However, the biggest impact in genomics and in understanding microbial genomes and other species has been the use of new mathematical models to deal with, first, the tens and hundreds of thousands and then the millions and tens of millions of sequences, to put together the picture of whole genomes.

MICROBIAL SEQUENCING

We decided to apply this new model, or assembly algorithm, dubbed the "Whole Genome Shotgun Strategy" in 1995 to *Hemophilus influenzae* as a test organism. It took less than a year, using the facility we had at the Institute for Genomic Research (TIGR). In 1995 we published the first complete genome map of a nonviral organism in *Science* (Smith et al., 1995).

A few years earlier we had published the complete sequence of smallpox (Massung et al., 1993), but it was orders of magnitude smaller. Observed in this sequence was something that has been found in every pathogen sequenced since—a finding that has changed our view of evolution. In front of a large number of the genes—basically all those coding for surface molecules (lipopolysaccharide biosynthesis)—there are tetrameric repeats that are preprogrammed evolutionary changes. Thus, every 10,000 or so replications there is slippage that changes the downstream reading frame, putting stop codons on the genes—that is, knocking them out. This is an important finding in understanding emerging infections or biological warfare agents. For example, a number of companies have tried to use the influenza sequence to develop new vaccines but have ignored the signals of the proteins designed to evolve very rapidly in real time. This resulted in no new stable vaccines because the virus changed too much in too short a period for the vaccine to be effective.

Currently, we are in an exponential growth phase in microbial sequencing. The first two microbes were sequenced in 1995, and four were sequenced in 1996 and 1997. To date, more than 30 have been completed. The American Society for Microbiology has indicated that the next phase will include the sequencing of more than 500 microbes. Sequencing of biological warfare pathogens is likely to increase over the next few years. For comparative genomic purposes, having a number of sequences complete is orders of magnitude more valuable than having just one complete.

Sequencing of *H. influenzae* uncovered a number of genes that are completely new to biology. Initially it was thought that this was due to a paucity of data in the databases. However, sequencing of subsequent genomes has revealed that, on average, roughly half of the genes in each genome are new, at least from certain biological perspectives. Some of these genes are highly conserved, but many in each species are novel, posing a real challenge to biology.

Before the first two genomes were sequenced—one a gram-positive organism and the other a gram-negative organism (with tremendous overlap)—we mistakenly thought that the gene pool on this planet was remarkably small. Most of the genes were of unknown function, however, and that changed the view of genetic diversity. The new data have not changed this view. Most likely, the percentage of new genes will remain the same, even after 500 organisms have been sequenced. Moreover, many techniques have been developed to try to identify genomes based on one or two sequences, but in view of the large percentage of previously unknown genes in each genome, it is unlikely that this approach will be effective. As we go forward in evolution, we see a great deal of gene duplication, increased specificity, and significant variation on a theme.

At TIGR we have spent a lot of time trying to characterize a simple organism, *Mycoplasma genitalium*, a species that might be a causative agent of urethritis. It is questionable whether it is a human pathogen or not. With 470 genes it has the smallest genome of any species known to be a self-replicating organism. Just how many of those are necessary for life in a rich growth environment is an important question.

Hutchison et al. (1999) developed a method to conduct whole-organism transposon mutagenesis. Having the complete genome sequence allows one to incorporate transposons in and then look to see where they insert in the genome. Because one can sequence off the transposon, their precise location can be determined. This approach has been aided by the sequencing of a second bacterium, *Mycoplasma pneumoniae*. It turns out that the entire *M. genitalium* genome is contained in the *M. pneumoniae* genome; however, *M. pneumoniae* has 200 more genes than *M. genitalium*. This finding provided a way to test evolutionary hypothesis, with the assumption that because the 200 genes in *M. pneumoniae* are not required for normal life and replication, we should be able to knock those genes out with less likelihood of the organism dying than if genes of *M. genitalium* were randomly removed.

The resulting transposon maps showed that a cell could survive only if the transposon did not insert in the middle of an essential gene and kill the organism. If the transposon inserted in a gene does not prove lethal, the gene is deemed to be nonessential. Eventually, it was found that 300 of the 470 genes could not be knocked out without killing the cell. In the next stage of the experiment, plans were made to create a synthetic organism, a step that raised some ethical concerns at TIGR.

Earlier at the National Institutes of Health (NIH), debate began when a laboratory started sequencing the smallpox genome. The concern was that, if the smallpox genome sequence was published, it would be akin to publishing a blueprint to a bomb because eventually anybody with a molecular biology tool kit would be able to synthesize smallpox. This is not far from the truth. It would be relatively trivial to synthesize and replicate the smallpox virus or even modify vaccinia. In short order, genomes will be synthesizable from scratch. Consequently, it will be critical to have the complete genetic sequence of every possible potentially emerging pathogen and potential biological warfare agent because only by examining the complete sequence can it be determined if it has been deliberately modified.

Even though our focus is often on microbes, plants also could be important targets of biological warfare. Fifty percent of the world's food production is covered by just three different species. With funding from the National Science Foundation, TIGR is in the process of sequencing the rice genome and the *Arabidopsis* genome, which are models for about 170,000 species. TIGR has just finished the first chromosome and has discovered many features that play key roles in evolution. A surprising finding is that as we go up the evolutionary tree, genes have a lot more introns, and these DNA play key regulatory roles.

SEQUENCING THE MALARIA GENOME

Malaria is a worldwide threat to civilian and military populations. The U.S. Department of Defense (DOD) estimates that if U.S. troops were deployed to some of the drug-resistant malaria regions of the world, there could be 20 to 30 percent casualties just from mosquito bites infecting soldiers with drug-resistant malaria. As a result, DOD, NIH, and some private foundations have funded TIGR and other organizations to sequence malaria genomes. For years many people thought that the malaria genome was undecipherable and could not be sequenced because of the high AT content. Although most clones could not even survive in *Escherichia coli*, we decided to try a whole genome shotgun method with small clones. This method worked well.

Some unique methods were used—such as DNA restriction digest of single molecules of single chromosomes—to correctly verify the assembly of the malaria genome. Moreover, even though this is a eukaryote, genes at both ends of the telomeres appear to play the same role that some of these sequences play in the bacteria—that is, they lead to constant antigen variation and evolution of the malaria species. This is why these organisms evolve rapidly, overcoming any attempted drug interventions.

SEQUENCING TUBERCULOSIS

In 1999 we finished sequencing the Oshkosh strain of tuberculosis, a highly virulent and emerging virulent strain that originated with an individual working at the Oshkosh clothing factory in rural Tennessee. Fortunately, the strain was drug sensitive. Comparing the genome of this strain to a laboratory strain revealed that it had more genes. In addition, some of the genes in the laboratory strains, which are still infectious, may have been spliced out by transposons inserting near each other in the genome. We now have the tools to understand for the first time why this new strain is so infective. These tools also allowed us to realize that this probably was not a new strain but rather a re-emergence of an ancient strain. Tuberculosis used to be much more infective than it appears to be now. This pattern could also occur with smallpox.

BIOCOMPUTING: THE MAJOR TOOL OF SEQUENCING

The scale of sequencing has changed tremendously in the past year with the introduction of a completely automated sequencing instrument from Applied Biosystems. The system uses capillaries, 96 at a time, through which the DNA flows into a cuvette. A laser then shines through the cuvette, activating all the dye simultaneously, giving us 100-fold more sensitivity and throughput than we had before. Celera has 300 high-throughput sequencing machines in a new facility. These machines will allow Celera to do 200,000 sequencing reactions per day, generating 2 billion base pairs of sequence every month.

Using these tools, Celera is concentrating on four key genomes—the human genome, the mouse genome, an insect genome, and the rice genome. It is hoped that by studying mutations in humans we can understand the susceptibility of humans to certain pathogens.

To put this in perspective, it took us a year to do the *H. influenzae* genome, which required the sequencing of approximately 25,000 clones, amounting to roughly 12 pathogens per 24-hour rate for complete genome sequencing. There are numerous other possible applications of this technology, including sampling environments to detect and culture microbial agents. The ability to run 200,000 samples a day completely alters what can be accomplished.

For example, in an emergency one could collect samples from an individual and within 24 hours have a pathogen's complete genomic sequence, work through the assembly algorithms, and run comparisons to databases. In this scenario many people in genomics have been concerned about how many base pairs could be analyzed. Celera has been concentrating on the computational side of this problem and is collaborating with COMPAQ Computers to build the second-largest supercomputer ever constructed.

Eventually, we will have more than 1,200 interconnected EV6 alpha processors. Although it is complicated, even with the best algorithms, to compare sequences to each other, just one of these new alpha chips can do over 250 billion of these pairwise comparisons an hour. We need this power just to deal with the daily throughput. These computers each have 20 to 30 gigabytes of RAM (random access memory). If there were a terabyte RAM machine at this time, we would be one of the first to purchase it because dealing with the computational side is very important.

Over the next 18 months, we plan to complete the genome sequence of the human. This requires the sequencing of 70 million clones, or the equivalent of 3,000 complete genomes the size of *H. influenzae.*

In terms of human variation, we all differ from one another in about 3 million letters of genetic code. Although these differences can lead to diseases, our sequencing of the complete genome of five people will provide a database of 20 million single nucleotide polymorphisms (SNPs) over the short term. Others have calculated that this database will represent 80 percent of the abundant polymorphisms in the human population, providing a tremendous resource for understanding the complexity of human diseases. A potential application of this in the next decade is in pharmacogenetics, as the pharmaceutical industry moves to use patient segmentation based on SNPs for drug trials.

SUMMARY

All of the information that is being generated by these sequence data needs to be organized and processed in order to be useful. To understand the structure of genomes, we need to be able to overlay the mouse genome on top of the human genome and those of other species. The supercomputers available today are woefully inadequate for this task. Biology is moving past the forefront of computing, requiring the development of new computational powers. However, there is a big difference between having the information and understanding it. We still must rely on the fundamentals of developmental biology to understand DNA repair, the cell cycle, and gene-gene and gene-environment interactions. Without the ability to process all of the data emerging from these sequencing efforts, we will not be able to make progress in understanding and curing complex multigenic diseases such as cancer. And we will not be able to understand human variability in response to infectious agents.

REFERENCES

Hutchison, C. A., S. N. Peterson, S. R. Gill, R. T. Cline, O. White, C. M. Fraser, H. O. Smith, and J. C. Venter. 1999. Global transposon mutagenesis and a minimal Mycoplasma genome. Science, 286(5447):2165-2169.

Massung, R. F., J. J. Esposito, L. I. Liu, J. Qi, T. R. Utterback, J. C. Knight, L. Aubin, T. E. Yuran, J. M. Parsons, V. N. Loparev, et al. 1993. Potential virulence determinants in terminal regions of variola smallpox virus genome. Nature, 366(6457):748-751.

Smith, H. O., J. F. Tomb, B. A. Dougherty, R. D. Fleischmann, and J. C. Venter. 1995. Frequency and distribution of DNA uptake signal sequences in the Haemophilus influenzae Rd genome. Science, 269(5223):538-540.

23

Summary and Next Steps

Scott P. Layne, Tony J. Beugelsdijk, and C. Kumar N. Patel

Scientists have fast computers, database technologies, and Internet connections (i.e., powerful "digital" firepower) for analyzing enormous problems in medicine and biology. Many important research efforts, however, currently are limited by the ability to carry out vast numbers of experimental tests in the laboratory. Solving such problems requires armies of laboratory technicians or sheer "physical" firepower. The critical gaps in digital versus physical firepower are at the heart of many martinet problems facing society, such as the following:

- Fighting deadly infectious diseases: influenza A epidemics; drug-resistant tuberculosis; HIV/AIDS drugs and vaccines.
- Ensuring safe food: reducing risks; detecting pathogens and toxins; investigating outbreaks.
- Bioterrorism and biowarfare: preventing attacks; characterizing agents; minimizing aftermaths; directing responses.
- Human genetics and molecular medicine: predicting cancers; diagnosing diseases; tailoring medications.

INTERMEDIATE SCALE

The scientific community needs new intermediate-scale resources with the firepower comparable to hundreds of humans. For the first time in scientific history a critical number of scientific disciplines and available technologies can be brought together to create such resources and level the playing field. The colloquium placed such needs in the forefront and then examined how available technologies can be brought together to fill critical gaps. The approach recognized that new tools, technologies, and standards from one scientific discipline can be carried over to others.

FEASIBILITY

Building new high-throughput laboratories is scientifically and technically feasible. Such intermediate-scale resources would take advantage of existing know-how, be operational within a few years, and enable new solutions to grand challenge problems.

OBSTACLES

The U.S. government is not geared to meet the twenth-first-century demands of intermediate-scale science. Today's funding programs can be divided into three scales: (1) small, (2) intermediate, and (3) large. Funding mechanisms are available for small- and large-scale efforts but not for intermediate-scale efforts.

NEEDS

The U.S. government needs to establish intermediate-scale funding programs to tackle critical problems faced by science and society. The goal of the colloquium was to examine these needs and offer policy makers education and guidance for creating key programs.

FIRST APPLICATION

A global laboratory against influenza may offer the best opportunity for moving forward. Systems are in place for collecting 170,000 samples per year and for manufacturing vaccines to reduce the impact of this evolving pathogen. The deadly outbreak of avian influenza in Hong Kong in 1997 and sequencing of the 1918 pandemic strain provide further justifications for moving forward.

APPENDIXES

APPENDIX A

CONTRIBUTORS

Ken Alibek, M.D., Ph.D.
Chief Scientist
Hadron, Inc.
Alexandria, Virginia

Roy M. Anderson, Ph.D.
Director and Professor
Wellcome Trust Center for the Epidemiology of Infectious Diseases
University of Oxford
Oxford, England

John C. Bailar III, M.D., Ph.D.
Professor and Chair
Department of Health Studies
University of Chicago
Chicago, Illinois

Tony J. Beugelsdijk, Ph.D., M.B.A.
Team Leader
Applied Robotics and Automation Group
Engineering Sciences and Applications Division
Los Alamos National Laboratory
Los Alamos, New Mexico

Hong Cai, Ph.D.
Staff Scientist
Life Sciences Division
Los Alamos National Laboratory
Los Alamos, New Mexico

Ashton Carter, Ph.D.
Professor
John F. Kennedy School of Government
Harvard University
Cambridge, Massachusetts

Nancy J. Cox, Ph.D.
Chief
Influenza Branch
Centers for Disease Control and Prevention
Atlanta, Georgia

T. Gregory Dewey, Ph.D.
Professor
Keck Graduate Institute of Applied Life Sciences
Claremont, California

David J. Galas, Ph.D.
Chief Academic Officer and Professor
Keck Graduate Institute of Applied Life Sciences
Claremont, California

Robert C. Habbersett, B.A.
Staff Scientist
Life Sciences Division
Los Alamos National Laboratory
Los Alamos, New Mexico

James M. Hughes, M.D.
Director
National Center for Infectious Diseases
Centers for Disease Control and Prevention
Atlanta, Georgia

James H. Jett, Ph.D.
Director
National Flow Cytometry Resource
Los Alamos National Laboratory
Los Alamos, New Mexico

Richard A. Keller, Ph.D.
Staff Scientist
Chemical Sciences and Technology Division
Los Alamos National Laboratory
Los Alamos, New Mexico

Gary W. Kramer, Ph.D.
Staff Scientist
National Institute of Standards and Technology
Gaithersburg, Maryland

Erica J. Larson, Ph.D.
Staff Scientist
Engineering Sciences and Applications
Los Alamos National Laboratory
Los Alamos, New Mexico

Scott P. Layne, M.D.
Associate Professor
Department of Epidemiology
School of Public Health
University of California, Los Angeles

David J. Lipman, M.D.
Director
National Center for Biotechnology Information
National Library of Medicine
Bethesda, Maryland

Babetta L. Marrone, Ph.D.
Staff Scientist
Life Sciences Division
Los Alamos National Laboratory
Los Alamos, New Mexico

Susan E. Maslanka, Ph.D.
Chief
National Botulism Surveillance and Investigation Laboratory
Foodborne and Diarrheal Diseases Laboratory Section
Centers for Disease Control and Prevention
Atlanta, Georgia

Randall S. Murch, Ph.D.
Deputy Assistant Director
Laboratory Division
Federal Bureau of Investigation
Washington, D.C.

Gary J. Nabel, M.D.
Professor of Medicine
Howard Hughes Medical Institute
University of Michigan
Ann Arbor, Michigan

John P. Nolan, Ph.D.
Staff Scientist
Life Sciences Division
Los Alamos National Laboratory
Los Alamos, New Mexico

Ariel Pablos-Mendez, M.D., M.P.H.
Assistant Professor
Division of General Medicine
Columbia College of Physicians and Surgeons
New York, New York

COL Gerald W. Parker, D.V.M., Ph.D.
Commander
U.S. Army Medical Research Institute of Infectious Diseases
Frederick, Maryland

C. Kumar N. Patel, Ph.D.
Vice Chancellor, Research
University of California, Los Angeles

William C. Patrick III
President
BioThreats Assessment
Frederick, Maryland

Joseph V. Rodricks, Ph.D.
Managing Director
The Life Sciences Consultancy
Washington, D.C.

Gary K. Schoolnik, M.D.
Professor
Department of Microbiology and Immunology
Stanford University
Palo Alto, California

Jeremy Sobel, M.D.
Medical Epidemiologist
Foodborne Diseases Epidemiology Section
Centers for Disease Control and Prevention
Atlanta, Georgia

Xuedong Song, Ph.D.
Staff Scientist
Life Sciences Division
Los Alamos National Laboratory
Los Alamos, New Mexico

Torsten A. Staab, M.S., F.H.
Staff Scientist
Applied Robotics and Automation Group
Engineering Sciences and Applications Division
Los Alamos National Laboratory
Los Alamos, New Mexico

Bala Swaminathan, Ph.D.
Chief
Foodborne and Diarrheal Diseases Laboratory Section
Centers for Disease Control and Prevention
Atlanta, Georgia

Basil Swanson, Ph.D.
Laboratory Fellow
Los Alamos National Laboratory
Los Alamos, New Mexico

Jeffery K. Taubenberger, M.D., Ph.D.
Chief
Division of Molecular Pathology
Armed Forces Institute of Pathology
Washington, D.C.

J. Craig Venter, Ph.D.
President
Celera Genomics
Rockville, Maryland

Paul S. White, Ph.D.
Staff Scientist
Life Sciences Division
Los Alamos National Laboratory
Los Alamos, New Mexico

Michael A. Wilson, Ph.D.
Postdoctoral Fellow
Department of Microbiology and Immunology
Stanford University
Palo Alto, California

Gerald Zirnstein, Ph.D.
Research Microbiologist
Helicobacter Antimicrobial Resistance Laboratory
Foodborne and Diarrheal Diseases Laboratory Section
Centers for Disease Control and Prevention
Atlanta, Georgia

APPENDIX **B**

AUTOMATION IN THREAT REDUCTION AND INFECTIOUS DISEASE RESEARCH

NEEDS AND NEW DIRECTIONS

Agenda of the April 1999 Colloquium

THURSDAY, APRIL 29, 1999

Session 1	**Moderator: C. Kumar N. Patel**
Welcome & Introduction 8:00 - 8:30 a.m.	C. Kumar N. Patel, University of California, Los Angeles, Calif. "Building an Intermediate-Scale Research Infrastructure to Tackle Big Problems"
Infectious Diseases 8:30 - 10:30 a.m.	James M. Hughes, Centers for Disease Control and Prevention, Atlanta, Ga. "Addressing Emerging Infectious Diseases, Food Safety, and Bioterrorism: Common Themes"
	Jeffery K. Taubenberger, Armed Forces Institute of Pathology, Washington, D.C. "Sequencing Influenza A from the 1918 Pandemic, Investigating Its Virulence, and Predicting Future Outbreaks"
	Nancy J. Cox, Centers for Disease Control and Prevention, Atlanta, Ga. "Expanding the Worldwide Influenza Surveillance System and Improving the Selection of Strains for Vaccines"

	Roy M. Anderson, University of Oxford, Oxford, U.K. "The Interplay of Mathematical Models and Laboratory Experiments in Infectious Disease Research"
Infectious Diseases 11:00 a.m. - 12:30 p.m.	Ariel Pablos-Mendez, Columbia College of Physicians, New York, N.Y. "Next Steps in the Global Surveillance for Antituberculosis Drug Resistance"
	Gary K. Schoolnik, Stanford University, Palo Alto, Calif. "Functional Genomics by Microarray Hybridization and the Drug-Discovery Process: The INH Mycobacterium Tuberculosis Paradigm"
	Scott P. Layne, University of California, Los Angeles "Laboratory Firepower: Batch Science via the Internet"
Questions & Answers 12:30 p.m. - 1:00 p.m.	Auditorium
Session 2	**Moderator: Scott P. Layne**
Infectious Diseases/ Technologies 2:00 - 3:30 p.m.	David J. Lipman, National Library of Medicine, Bethesda, Md. "The I/O of High-Throughput Biology: Experience of the National Center for Biotechnology Information"
	Peter L. Nara, Biological Mimetics Inc., Frederick, Md. "Key Assays and Data for HIV Research and Development Efforts"
	Gary J. Nabel, University of Michigan, Ann Arbor "Developing Therapies and Vaccines for HIV/AIDS and Emerging Infectious Diseases"

Food Supply/ Technologies 4:00 - 6:00 p.m.	John C. Bailar III, University of Chicago, Ill. "Ensuring Safe Food: An Organizational Perspective" Joseph Rodricks, Life Sciences Consultancy, Washington, D.C. "Ensuring Safe Food: Risk Assessments and Data Needs" Bala Swaminathan, Centers for Disease Control and Prevention, Atlanta, Ga. "Food Pathogen Diagnostics from Public Health and Bioterrorism-Preparedness Standpoints" Tony J. Beugelsdijk, Los Alamos National Laboratory, Los Alamos, N. Mex. "Standard Laboratory Modules for Batch Science via the Internet"
Questions & Answers 6:00 - 6:30 p.m.	Auditorium

FRIDAY, APRIL 30, 1999

Session 3	**Moderator: Maurice Hilleman**
Biowarfare & Bioterrorism 8:00 - 10:00 a.m..	William C. Patrick III, BioThreats Assessment, Frederick, Md. "Biological Warfare Scenarios" Ken Alibek, Hadron Inc., Annandale, Va. "Scenarios for Biological Weapons Effects" Gerald Parker, U.S. Army Medical Research Institute of Infectious Diseases, Frederick, Md. "Front-Line Responses to Biowarfare and Bioterrorism Pathogens" Ashton B. Carter, Harvard University, Cambridge, Mass. "National Innovations to Combat Catastrophic Terrorism"

Biowarfare & Bioterrorism/ Technologies 10:30 a.m. - 12:00 p.m.	Thomas Marr, Genomica Corporation, Boulder, Colo. "New Technologies for High-Throughput Genetic Characterization and Functional Analysis"
	James H. Jett, Los Alamos National Laboratory, Los Alamos, N. Mex. "Flow Cytometric Detection Methods for High-Throughput Research"
	Stephen S. Morse, Defense Advanced Research Projects Agency, Arlington, Va. "Developing Advanced Diagnostics for Detecting Pathogens: Applications in High-Throughput Research"
Questions & Answers 12:00 - 12:30 p.m.	Auditorium
Session 4	**Moderator: Tony J. Beugelsdijk**
Further Applications/ Technologies 1:30 - 3:30 p.m.	David J. Galas, Keck Graduate Institute of Applied Life Sciences, Claremont, Calif. "Integrating Key Technologies for the Future of the Life Sciences"
	Leena Peltonen, University of California, Los Angeles "Human Genetics and Molecular Medicine: How Technologies Provide New Views on Health and Disease"
	Gary W. Kramer, National Institute of Standards and Technology, Gaithersburg, Md. "New Standards and Approaches for Integrating Instruments into Laboratory Automation Systems"
	Randall Murch, Federal Bureau of Investigation, Washington, D.C. "The FBI Laboratory Bioterrorism Program"

Further Applications/ Technologies 4:00 - 5:30 p.m.	J. Michael Ramsey, Oak Ridge National Laboratory, Oak Ridge, Tenn. "Lab-on-a-Chip Devices for High-Throughput Biochemical Experimentation"
	Willem Stemmer, Maxygen Inc., Redwood City, Calif. "DNA Shuffling, DNA Vaccines, and Rapid-Design Strategies"
	J. Craig Venter, Celera Genomics, Rockville, Md. "High-Throughput Sequencing, Information Generation, and the Future of Biology"
Questions & Answers 5:30 - 6:00 p.m.	Auditorium

INDEX

A

C

D

E

F

G

H

I

M

N

O

P

Q

R

R
858
.A2
F55

2001